SCIENCE ASKEW

A LIGHT-HEARTED LOOK AT THE SCIENTIFIC WORLD

NASA
USDA
TORY PETERSON
DEMO LAB TODA
PULLEYS, FORCES
CENTER OF GRA
ZZZZZZ
JCI

Science Askew

A Light-Hearted Look at the Scientific World

Donald E. Simanek

and

John C. Holden

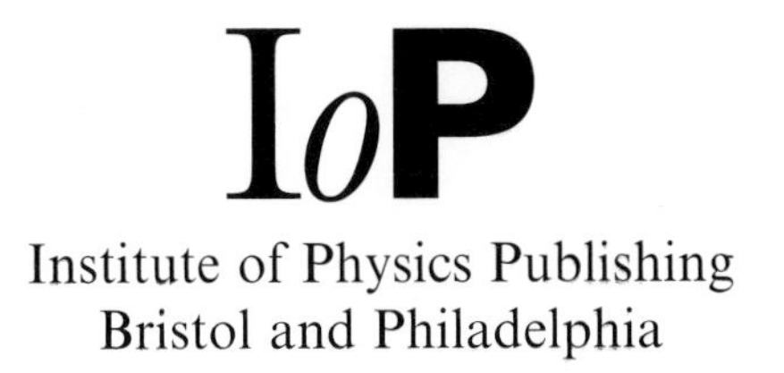

Institute of Physics Publishing
Bristol and Philadelphia

British Library Cataloguing-in-Publication Data
A catalogue record for this book is available from the British Library.

ISBN 0 7503 0714 5

Library of Congress Cataloging-in-Publication Data are available

Commissioning Editor: Nicki Dennis
Production Editor: Simon Laurenson
Production Control: Sarah Plenty
Cover Design: Frédérique Swist
Marketing Executive: Laura Serratrice

Published by Institute of Physics Publishing, wholly owned by The Institute of Physics, London

Institute of Physics, Dirac House, Temple Back, Bristol BS1 6BE, UK

US Office: Institute of Physics Publishing, The Public Ledger Building, Suite 1035, 150 South Independence Mall West, Philadelphia, PA 19106, USA

Typeset by Academic + Technical Typesetting, Bristol
Printed in the UK by Bookcraft, Midsomer Norton, Somerset

Science Askew

By Donald Simanek and John Holden

An Almanac of Scientific Ephemera.

Being a light-hearted look at *SCIENCE*, a compendium of levity on *MATTERS OF GRAVITY*, an inveſtigation into the murky realms of the duſty laboratories and minds of scientiſts in all fields of the *NATURAL SCIENCES* (and *MATHEMATICS*, too), compiled for the edification of diverſe and sundry addicts of the abſurd, aficionados of the abstruſe and connoiſſeurs of the contrived, being soundly baſed upon the work of *INGENIOUS* inveſtigators of the *MYSTERIES* of *NATURE* in many conſiderable parts of the *WORLD*, & including digreſſions into philoſophie and religion to boot. Theſe eſſays are deſigned to be inſructive and illuminating as well as entertaining, suitable for any student with an *INQUIRING MIND*, impatient with staid and stuffy reſearch inveſtigations reported in scientific journals.

Inſtitute of Phyſics Publiſhing, Briſtol and Philadelphia, MMII

CONTENTS

INTRODUCTION

No mind is thoroughly well organized that is deficient in a sense of humor.

Samuel Taylor Coleridge (1772–1834)
English poet, critic and philosopher

I hate quotations. Tell me what you know.

Ralph Waldo Emerson (1803–1882) U.S. essayist and poet

Science is a serious discipline. Sometimes it seems **too** serious. The mind of the science student is so stuffed with technical words, laws, principles, and theory that sometimes things begin to leak out. Even seasoned scientists, their minds saturated with mathematics and logic, find release and recreation in bouts of frivolous illogic, fantasy, word-play, parody, and sophomoric humor.

This book is a tribute to that spin-off of the scientific enterprise, a scrapbook of levity on matters of gravity. It's a compendium of perversions and twists of the rigid discipline of scientific thinking. Such flights of fancy are seldom seen in serious and stuffy research journals. They are part of the oral tradition, or passed from hand to hand on mimeographed, dittoed, or xerographic copies. Today, much of this sort of humor resides on the Internet, that world-girding pothole-filled highway of information and misinformation. We include examples from these sources without apology (except to the original authors, who are often unknown and therefore unheralded). Their uninhibited humor illustrates the creative and inventive mind at play.

The reader will find here a generous helping of our own humor which we've concocted over the years during odd moments (and even moments, too) stolen from the more mundane and serious work one must do to make a living as a scientist.

A copious cornucopia of John Holden's drawings, illusions, and cartoons enliven the proceedings. They are easily distinguished by their style (and the initials JCH) from pictures we've swiped from other sources.

A warning: Do not read this book cover to cover, and certainly not at one sitting. If one chapter is not to your liking move quickly to the next; you may like it even less.

ACKNOWLEDGMENTS

Some of the illustrations have appeared in previously published books, magazines, and newsletters. We would like to thank all authors and publishers for permissions to reprint these illustrations.

Pages 66, 78, 251, 257, 268, 271, 272, and 275 are from *Creation/Evolution Satiricon* (Bookmaker, 1987) by R. S. Dietz and J. C. Holden.

Pages 112, 283, and 294 are from *Now That You Are Here: Methow Valley Guidebook* (Bookmaker, 1986) by J. C. Holden, R. W. Hult, and F. Brewster.

Pages 153, 242, 246, 267, and 293 are from *The Id of the Squid* (Compass Publications, 1970) by A. E. Benthic.

Page 154 is from *Grungy George and Sloppy Sally* (Vantage Press, 1993) by H. B. Stewart.

Page 265 from *Creationism's Geologic Time Scale* (American Scientist, March–April, 1998) by D. U. Wise.

Page 201 from the newsletter *Science Frontiers* (the *Sourcebook Project*, May–June, 1997).

Page 38 from *Graphic Solutions to Problems of Plumacy* EOS, Trans. American Geophysical Union, **68**, 7 (July, 1977) by J. C. Holden and P. R. Vogt.

Pages 40 and 42 are from *The Journal of Irreproducible Results* **22**, 2 (July, 1976).

Pages 55 and 59 are from *The Journal of Irreproducible Results* **20**, 4 (July, 1979).

A quote note

We couldn't resist sprinkling our favorite quotations throughout this collection. It's helpful to know who first uttered or wrote a quotable quote, something about the person, when he or she lived, and when it was said. So we have tried to indicate those facts where we've been able to track them down. Unfortunately, some quotes, too good to omit, came from sources which provided no more than a name, and we were unable to identify the author. To those folks (if they are still living) we apologize, and welcome additional information from anyone who knows more about them.

Quote collections are notoriously prone to errors. Any clever saying someone invents which is catchy enough to be widely quoted was very likely said by someone earlier. When a celebrity utters a popular, commonly known saying, that person is forever after credited with it. Consider the often-repeated comment of Newton in a letter to Hooke, 5 Feb. 1676. (*Corres* I, 416):

> *If I have seen further [than others] it is by standing upon the shoulders of giants.*
>
> Sir Isaac Newton (1642–1727)
> English mathematician and physicist

Compare:

> *Pygmies placed on the shoulders of giants see more than the giants themselves.*
>
> Lucan (Marcus Annaeus Lucanus) (39–65)
> Roman poet, born in Spain

Much later, in the 12th century Bernard of Chartres observes: *Nos esse quasi nanos gigantum humeris insidientes. (We are as dwarfs sitting on the shoulders of giants.)* Robert Merton did a comprehensive study of this quote, finding 26 other authors who used it between the time of Bernard and Newton. And that's only those persons who achieved some literary recognition.

We gave up trying to identify the original originator of:

> *Truth arises from error more easily than from confusion.*

But we don't hesitate to offer these variants, for future quote collectors:

> *Truth arises from confusion and doubt more easily than from misplaced and unwarranted certainty and belief.*
>
> Donald E. Simanek

> *If I have seen further than others it is by standing upon the shoulders of others who stood upon the shoulders of giants—and using binoculars.*

It's shoulders all the way up.

1

[Science is] a great game. It is inspiring and refreshing. The playing field is the universe itself.

Isidor Isaac Rabi (1898–1988) U.S. physicist. Nobel Prize 1944

More than ever, the creation of the ridiculous is almost impossible because of the competition it receives from reality.

Robert A. Baker (1937–) U.S. author

A man must have a certain amount of intelligent ignorance to get anywhere.

Charles Franklin Kettering (1876–1958)
U.S. engineer and inventor

Critics, always critics

Even scientists and mathematicians must endure critics. Some people just didn't trust mathematics.

> *The good Christian should beware of mathematicians and all those who make empty prophecies. The danger already exists that the mathematicians have made a covenant with the devil to darken the spirit and to confine man in the bonds of Hell.*
>
> Saint Augustine (354–430)

Augustine was criticizing astrologers, who used mathematics in constructing horoscopes.

Physicists were suspect also. In 1163 Pope Alexander III forbade "the study of physics or the laws of the world," to all ecclesiastics, and ordered that any person violating this rule "shall be avoided by all and excommunicated."

Galileo had many critics. One objected to Galileo's use of mathematics.

> *Mathematics is inadequate to describe the universe, since mathematics is an abstraction from natural phenomena. Also, mathematics may predict things which don't exist, or are impossible in nature.*
>
> Ludovico delle Colombe

The American writer and journalist Henry Louis Mencken expressed a healthy skepticism toward science.

> *Astronomers and physicists, dealing habitually with objects and quantities far beyond the reach of the senses, even with the aid of the most powerful aids that ingenuity has been able to devise, tend almost inevitably to fall into the ways of thinking of men dealing with objects and quantities that do not exist at all, e.g., theologians and metaphysicians. Thus their speculations tend almost inevitably to depart from the field of true science, which is that of precise observation, and to become mere soaring in the empyrean. The process works backward, too. That is to say, their reports of what they pretend actually to* see *are often very unreliable. It is thus no wonder that, of all men of science, they are the most given to flirting with theology. Nor is it remarkable that, in the popular belief, most astronomers end by losing their minds.*
>
> Henry Louis Mencken. Minority Report,
> H. L. Mencken's Notebooks, Knopf, 1956

WANTED—ON ANY TERMS

A REWARD IS OFFERED FOR INFORMATION LEADING TO THE ARREST OF **EDDY CURRENT**, charged with induction of an 18-year old coil, called Milli Henry, found half-choked, and also with the theft of valuable joules. This unrectified criminal armed with a carbon rod escaped from Weston Primary Cell, where he had been clapped in ions. The escape was planned in three phases. First he fused the electrolytes, then he climbed through a grid despite the impedance of wardens and finally ran to earth in a nearby magnetic field. He has been missing since Faraday. Watt seems most likely is that he stole an A.C. motor. This is of low capacity and he is expected to charge it for a megacycle and return ohm by a short circuit. He may offer series resistance and is a potential killer.

Not the intended message

A science teacher wanted to teach her 5th grade class a lesson about the health effects of liquor, so she performed an experiment using a glass of water, a glass of whiskey and two worms.

"Now, class, observe the worms closely," she said as she put a worm into the water. The worm wiggled about happily.

She put the second worm into the whiskey. It writhed as if in agony, and sank to the bottom, dead as a doornail.

"Now what lesson can we learn from this experiment?" she asked the class.

Little Johnny raised his hand and wisely observed, "Drink whiskey and you won't get worms."

Bedside manner

Sick patient: Doctor, am I at death's door?

Doctor: Don't worry, I'll pull you through.

Scientists never grow old, for the innocent pleasures of discovery will last them a lifetime.

Ernest Rutherford (1st Baron Rutherford of Nelson)
(1871–1937) English physicist, born in New Zealand.
Nobel Prize for Chemistry 1908

Work is the curse of the laboring class

A chemist was asked by his boss to do research on acetates. He refused, knowing that he who acetates is lost. Another chemist lost his job in adhesives research because he just couldn't stick to his work. Then there was the astronomer who lost his job in variable star research because he couldn't get Sirius, which reminds us of the worker in the orange juice factory who just couldn't concentrate.

Things are so bad these days that many an archaeologist's career is in ruins.

Objection, your Honor! This is speculation

The defense attorney in a murder trial was cross-examining a pathologist.

Attorney: Before you signed the death certificate, had you taken the pulse?

Coroner: No.

Attorney: Did you listen to the heart?

Coroner: No.

Attorney: Did you check for breathing?

Coroner: No.

Attorney: So, when you signed the death certificate, you weren't sure the man was dead, were you?

Coroner: Well, let me put it this way. The man's brain was sitting in a jar on my desk. But I guess it's possible he could be out there practicing law somewhere.

Symmetry

Seen on a poster for staff working in a hospital maternity unit: "Remember, the first 5 minutes of life are the most dangerous."

Comment added below: "The last 5 minutes are pretty dodgy too!"

Gilbert's salary theorem

A proof that engineers and scientists can never make as much money as administrators or salespersons.

1. Power = work/time.
2. Knowledge is power.

3. Therefore knowledge = work/time.
4. Time is money.
5. So knowledge = work/money.
6. And money = work/knowledge.
7. Therefore as knowledge $\rightarrow 0$, money $\rightarrow \infty$ regardless of the amount of work done.
8. Conclusion: The less you know, the more money you make.

Do it now!

- Statisticians probably do it.
- Algebraists do it in groups.
- (Logicians do it) OR [NOT (logicians do it)].
- Astronauts need space to do it.
- Professors do it absent-mindedly.
- Logicians do it symbolically.
- Sylvaculturists pine to do it.
- Mathematicians do it abstractly.
- Geneticists do it in their genes.
- Psychiatrists do it on couches.
- Engineers do it with precision.
- Cosmologists do it with a bang.
- Philosophers only think about doing it.

There are lots of other "do it" jokes, but we didn't want to overdo it.

A light breakfast

We seldom appreciate the role of food scientists in creating the form and substance of the things we eat. Now German food scientists have announced another breakthrough. They've found a way to trap so much air in waffle batter that the baked waffles are incredibly light—so light that you must pour on the syrup quickly to prevent them from floating up from the plate. This process will soon be marketed to consumers under the trade-name *Luftwaffles*.

Pupil dilation will do that

A fellow went to his optometrist optimistically and left misty optically.

Perceptive diagnosis

Another fellow went to see his doctor. The doc noticed right away that the man had a carrot stuck in one ear and a stalk of celery in the other. "Aha!" he said, "I see your problem. You aren't eating right."

A sweet reward

Business and industry fiercely compete to be the first with marketable ideas for products. We have even seen a number of cases of corporate espionage in the news lately.

One such case occurred in the candy industry. A spy sneaked into the Hershey chocolate factory, intent on filching technical secrets. While surreptitiously taking notes on the machinery and processes and trying to avoid detection he accidentally slipped on some freshly spilled chocolate and tumbled onto the conveyer belt. The machinery swallowed him up, coated him with chocolate, and dumped him out the other end wrapped up all bright and shiny as a giant Hershey Kiss. He was so humiliated that all he could say as he was being unwrapped was: "Curses, foiled again!"

It could have been worse. Consider the sad fate of the worker in the lensmaking factory who fell into the lens grinding machine and made a spectacle of himself.

Then there was the chemist who fell into the esterification kettle. He was saved, but was horribly butylated.

Endnotes

Never run after a bus, a woman, or a cosmological theory. There will always be another one in a few minutes.
John A. Wheeler (1911–) U.S. physicist

A chemical is a substance that:

An organic chemist turns into a foul odor.
An analytical chemist turns into a procedure.
A physical chemist turns into a straight line.
A biochemist turns into a helix.
A chemical engineer turns into a profit.

A chemist walked into a pharmacy and asked the pharmacist, "Do you have any acetylsalicylic acid?"

"You mean aspirin?" asked the pharmacist.

"That's it, I can never remember that word."

First, scientists patented living organisms. Now they are tailoring designer genes. With the sorry state of science education these days, students think that a gene splice is a mended pair of Levis. And when their denim pants shrink in the wash, they end up with recessive jeans.

When you reflect on the dismal state of the world, consider this. What can you expect of a universe that started out with nothing but a lot of hot gas?

Tell a man there are 300 billion stars in the universe and he'll believe you. Tell him a bench has wet paint and he'll have to touch it to be certain.

Some say that we have been visited by aliens from outer space. They descended on Washington, D.C., but left in disappointment. They were looking for signs of intelligent life on earth.

The most compelling evidence that intelligent civilizations exist elsewhere in the universe is the fact that none of them have tried to contact us.

2

Three Laws of Thermodynamics (paraphrased):

First Law: You can't get anything without working for it.

Second Law: The most you can accomplish by doing work is to break even.

Third Law: You can't even break even.

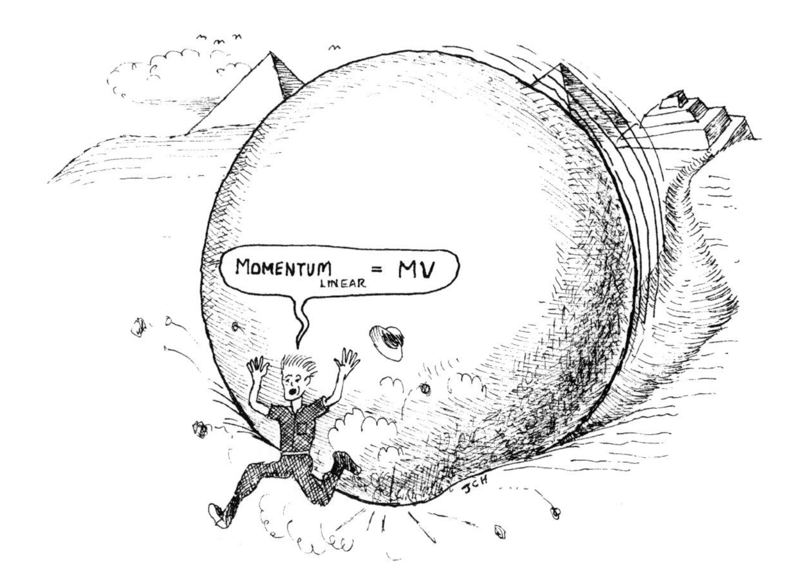

Rolling stones gather momentum.

Archimedes loses his principle.

Archimedes' lost principle: When a body is immersed in water, the phone rings.

Archimedes' displacement principle: Ringing telephones tend to displace bodies immersed in water.

The Principle of Virtual Work. Paperwork is inversely proportional to useful work.

Aristotle's rule of appliance maintenance. Vacuum cleaners break down because nature abhors a vacuum.

☞ Scientists calculate that interstellar matter in the universe consists of at least 1% dust, not counting that swept under rugs.

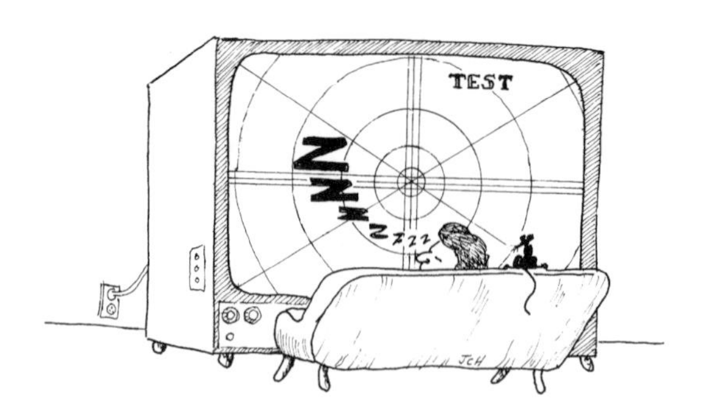

A body at rest tends to watch TV.

Phygg Newton's three laws of motion:

1. A body at rest tends to watch television.
2. A body acted upon by a force usually breaks.
3. For every human action there is an overreaction.

Euclid's entropy factory.

Euclid's entropy principle: Entropy tends to increase on horizontal flat surfaces.

Euclid's minimal principle: The shortest distance between two points is usually under construction.

Captain Hooke's law: Beware of crocodiles which go "tick-tock."

3

PHYSICS

Physics makes the world go 'round.[1]

The angst of physics

At the close of the 19th century many physicists confidently assumed that nothing fundamentally new remained be discovered.

> *It seems probable that the grand underlying principles of physical science have been firmly established and that further advances are to be sought chiefly in the rigorous application of these principles to all phenomena.*
>
> Albert Michelson (1852–1931)
> U.S. physicist. Nobel Prize 1907

Physicists were confident that if all aspects of any physical system were measured to perfect accuracy then physical laws could also be known perfectly and the future of that system could be mathematically predicted exactly. So all that was left for physicists to do was to improve measurements, pushing toward ever more decimal places of accuracy.

This smug complacency was shaken by the discovery of radioactivity, X-rays, and studies of light emission by matter. Still there was hope that the classical methods would make sense of it all.

Ernest Rutherford's experiments helped lay the foundations of modern atomic theory. He, like many other physicists of his time, assumed that the new phenomena would yield to analysis out of

[1] **Translation:** (1) The fact that the world is round and nearly spherical, rather than some other shape, is determined by laws of physics. (2) The earth's spin (spinning around its axis) and its revolution around the sun, are governed by physics principles.

which would come simple theories which could be understood without extraordinary mathematics.

If a piece of physics cannot be explained to a barmaid, then it is not a good piece of physics.

If your experiment needs statistics, then you ought to have done a better experiment.

Ernest Rutherford (1st Baron Rutherford of Nelson)
(1871–1937) English physicist, born in New Zealand.
Nobel Prize for Chemistry 1908

Rutherford's mistrust of statistical analysis seems naive today. But he was not alone in that opinion.

No effect that requires more than 10 percent accuracy in measurement is worth investigating.

Walther Nernst (1864–1941) German physicist, chemist.
Nobel Prize 1920

If results were not as good as one might like, the fault must be with the measurements. Where data and results are good, statistical methods aren't necessary.

The theorist's job is made difficult by subversive experimentalists bent on discovering data which doesn't fit theories. This is what experimentalists do, for it produces published papers. Each new physics journal brought reports of quantum phenomena which didn't fit classical physics models. Nature didn't seem neat and beautiful in the way those schooled in classical physics had come to expect. Understanding of the new data and theories didn't come easily, and many physicists expressed their frustration.

Physics is very muddled again at the moment; it is much too hard for me anyway, and I wish I were a movie comedian or something like that and had never heard anything about physics!

Wolfgang Pauli (1900–1958) Austrian physicist in the U.S.
Nobel Prize 1935. From a letter to R. Kronig, 25 May 1925

I do not like it, and I am sorry I ever had anything to do with it.

Erwin Schrödinger (1887–1961) Austrian physicist.
Nobel Prize 1933. Speaking of quantum mechanics

Those who are not shocked when they first come across quantum mechanics cannot possibly have understood it.

If anybody says he can think about quantum problems without getting giddy, that only shows he has not understood the first thing about them.

Niels Henrik David Bohr (1885–1962) Danish physicist

Heisenberg's uncertainty principle confounded naive notions of classical determinism. It showed how certain conjugate pairs of measurable quantities were related in such a way that improving the measurement precision of one variable necessarily worsened the precision of the some other one. Nature puts barriers in our way, limiting how much we can know. Every experiment designed to discredit or circumvent the uncertainty principle ended up supporting it. The comfortable determinism of classical physics was replaced by confusion and uncertainty.

To make things worse it seemed that on the atomic scale, nature behaved in an inherently statistical manner. Rutherford thought that statistical analysis was only a fudge for dealing with poor experiments and poor data. Now it seemed that nature itself was doing statistical fudging, preventing us from ever obtaining perfect simultaneous measurements of all aspects of systems.

There was nothing wrong with the old inference that if I know all about the present I can forecast the future exactly; the trouble was the impossibility of knowing the present. Once this is seen, the whole argument becomes obvious; but nobody saw it until Heisenberg.

Charles Galton Darwin (1887–1962) British physicist

We are in the ordinary position of scientists of having to be content with piecemeal improvements: we can make several things clearer, but we cannot make anything clear.

Frank Plumpton Ramsey (1903–1930)
English mathematician and philosopher

Albert Einstein couldn't accept that nature behaved this way. In an off-hand comment he said *Raffiniert ist der Herr Gott, aber boshaft ist er nicht.* "God is subtle, but not perverse."[2] Neither Einstein, nor anyone else, was able to make these unpleasant facts of nature go away. In discussions with Bohr, Einstein argued against the statistical nature of quantum mechanical laws, saying "I am convinced that He (God) does not play dice."

[2] According to Derek Price (1946) Einstein's own translation was "God is slick, but he ain't mean."

Albert is shocked to see God playing dice with Bohr.

Physicists no longer expected their theories to conform to naive notions of common sense.

In physics, instead of saying, I have explained such and such a phenomenon, one might say, I have determined causes for it the absurdity of which cannot be conclusively proved.

Georg Christoph Lichtenberg (1742–1799)
German physicist, philosopher, aphorist

But still every theory must connect in some way with experimental data, even though the connections may sometimes be tortuous and remote. The relation between theory and fact may be clarified by these two observations:

It is a capital mistake to theorize before one has data. Insensibly one begins to twist facts to suit theories, instead of theories to suit facts.

Arthur Conan Doyle (1859–1930)
British physician and novelist[3]

[3] In spite of this wise observation, Doyle believed in fairies and in spirit communication with the dead.

But are we sure of our observational facts? Scientific men are rather fond of saying pontifically that one ought to be quite sure of one's observational facts before embarking on theory. Fortunately those who give this advice do not practice what they preach. Observation and theory get on best when they are mixed together, both helping one another in the pursuit of truth. It is a good rule not to put overmuch confidence in a theory until it has been confirmed by observation. I hope I shall not shock the experimental physicists too much if I add that it is also a good rule not to put overmuch confidence in the observational results that are put forward ***until they have been confirmed by theory.***

Sir Arthur Eddington (1882–1944)
English astronomer and physicist

Existing theory, if it's well established, is never all wrong, and is the primary framework for posing questions for experimental observational test. Experimentation is seldom blind; it is motivated by theory.

It is the theory which decides what we can observe.

Albert Einstein (1879–1955)
Swiss–U.S. physicist Nobel Prize 1921

Physics theory has now become virtually incomprehensible to the layman. After all, the easy-to-comprehend problems of physics were solved early in history. So it shouldn't be surprising that there are no easy problems, nor easy answers left.

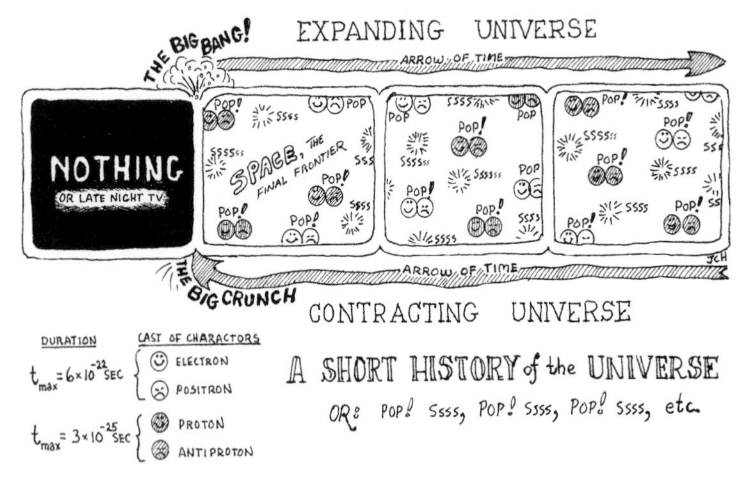

My own suspicion is that the universe is not only queerer than we suppose, but queerer than we can suppose.

J.B.S. Haldane (1892–1964) British geneticist and scientist.
On Being the Right Size in the (1928) book *Possible Worlds*

A shocking story

The children of the head of a prominent family wanted to give him a gift book of their family's history. The biographer they hired for the job was warned of one problem—Uncle Willie, the black sheep of the family, who had gone to the electric chair for murder.

The biographer assured them that there would be no problem: "I'll just say that Uncle Willie occupied a chair of applied electricity at one of our leading government institutions. He was attached to his position by the strongest of bonds. His death came as a true shock."

The theorist

When a theoretical physicist is asked, let us say, to calculate the stability of an ordinary four-legged table he quickly arrives at preliminary results which completely and rigorously treat the case of a one-legged table. He then treats, with wonderful elegance, the case of a table with an infinite number of legs. Then he will spend the rest of his life unsuccessfully trying to solve the ordinary problem of the table with an arbitrary, finite, number of legs.

And it has no moving parts

An engineer, a physicist, a mathematician, and a mystic were asked to name the greatest invention of all time.

The engineer chose fire, which gave humanity power over matter.

The physicist chose the wheel, which gave humanity the power over space.

The mathematician chose the alphabet, which gave humanity power over symbols.

The mystic chose the thermos bottle.

"Why a thermos bottle?" the others asked.

"Because the thermos keeps hot liquids hot in winter and cold liquids cold in summer."

"Yes—so what?"

"Think about it," said the mystic reverently. "That little bottle—how does it know?"

Just a special case

A physicist has completed a series of experiments, and has worked out an empirical equation that seems to explain his data. He asks a mathematician to look at it.

A week later, after taking sufficient time to consider the matter thoroughly, the mathematician reports that he can prove that the equation is invalid. By then, the physicist has used his equation to predict the results of further experiments, and he is getting excellent results. So he asks the mathematician to recheck his proof.

Another week goes by, and they meet once more. The mathematician reports that he has concluded, after exhaustive analysis, that the equation does work, "But only in the trivial case where the numbers are real and positive."

Merely a misplaced decimal point

The renowned cosmogonist Professor Bignumska, lecturing on the future of the universe, states that in about five billion years, according to theoretical calculations, the earth will fall into the sun in a fiery death. In the back of the auditorium a tremulous voice pipes up: "Excuse me, Professor, but h-h-how long did you say it would be?"

Professor Bignumska calmly replies, "About five billion years."

A sigh of relief is heard. "Whew! for a minute there, I thought you'd said five million."

That's what you call a robust theory

The experimentalist comes running excitedly into the theorist's office, waving a graph taken from his latest experiment. "Hmmm," says the theorist, "That's exactly where you'd expect to see that peak. Here's the reason." A long and thorough logical explanation follows. The experimentalist, who has been studying the graph, says "Whoops! This graph is upside down!" He fixes it. The theorist looks at the graph again and says, "Of course! You'd expect to see a dip in exactly that position. Here's the reason..."

☞ There are three kinds of robust theories: (1) Theories which turn out to be correct even in new situations never previously studied. (2) Theories which are correct even if their assumptions are untrue. (3) Theories which can accommodate even contradictory data.

A light pun

A rancher in Montana decided to retire and let his three sons operate his cattle ranch. He wanted more time to devote to his hobbies of photography and astronomy. His sons had quite a few innovative ideas to modernize the whole operation, making it more efficient and profitable. To signify its new identity, they sought a new name for the ranch. Their father suggested they call it "Focus." "Why?" his sons asked. "Because it's where the sons raise meat."

Oversupplied with the wrong half

The atomic bomb project at Oak Ridge, Tennessee, during World War II was top secret. This generated rumors of what might be going on there. One plausible-sounding rumor said that they were making front halves of horses, to be shipped to Washington, D.C., for completion of assembly.

What's Up?

The zenith's up,
The nadir's down,
And that's a fact
The world around.

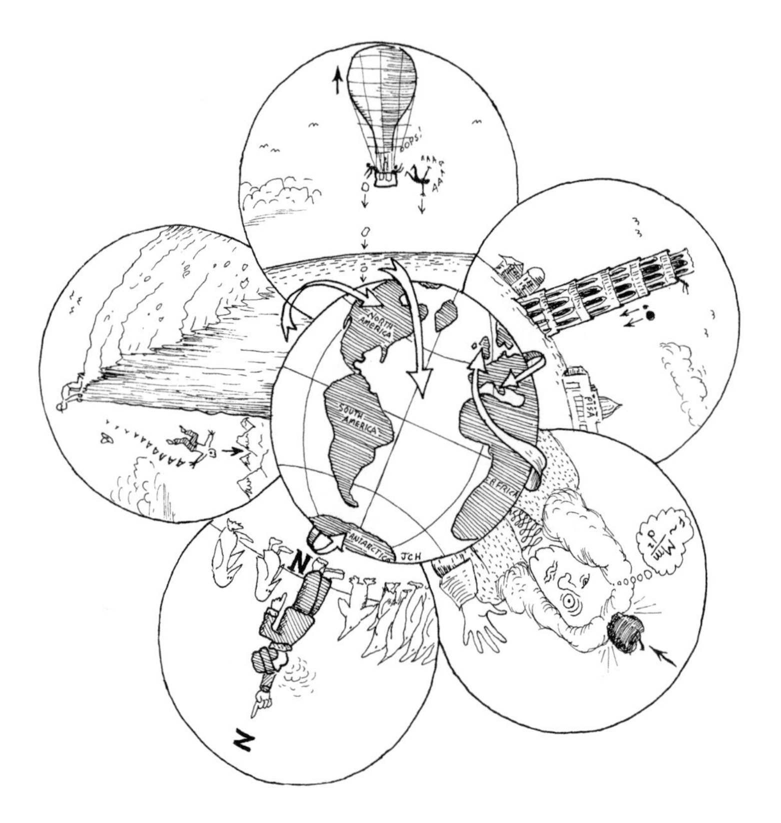

The relativity of "up"

Astronomers define the zenith as the point on the celestial sphere directly overhead of any observer wherever on earth the observer happens to be. The nadir is the point directly opposite, or the point straight down from where the observer is standing. This picture should make the concept clear.

The Physicist's Bill of Rights

We hold these postulates to be intuitively obvious: that all physicists were born equal (to a first approximation), and were endowed by their creator with certain indiscrete rights. Among these are a mean rest life, *n* degrees of freedom, and the pursuit of the physics secretary. We also reserve the following rights, which are invariant under all linear transformations:

1. To approximate all problems to ideal conditions.
2. To use order of magnitude calculations whenever necessary.
3. To dismiss any functions which diverge as "nasty," "badly behaved" and "unphysical."
4. To invoke the uncertainty principle when challenged by confused mathematicians, electrical engineers, and others.
5. When pressed by non-physicists for an explanation of (4) to mumble in a sneering voice about physically naive mathematicians.
6. To equate two sides of an equation which are dimensionally inconsistent with a suitable comment such as, "Well we are only interested in orders of magnitude anyway." See (2).
7. To justify shaky reasoning by observing that it does happen to give the correct answer.
8. To use plausibility arguments in place of proofs, and thenceforth to refer to these arguments as proofs.
9. To take on faith any principle which seems right but can't be proven.
10. When confronted with a question which requires any thought at all to reply, "Why not?"

Scraps from the laboratory floor

Theoretical physicist: A physicist whose existence is postulated to make the equations balance, but who is never actually observed in the laboratory.

Every experiment demonstrates something. If it doesn't demonstrate what you expected, it demonstrates something else.

Entropy requires no maintenance.

Seen on the door of an optics laboratory: "CAUTION! Do not look into laser with remaining good eye."

Absolute zero is real cool!

Student on meeting a physics professor: "Hi, Doc, what's new?" Professor: "e/h."

Banana physics

They are doing great things with vegetables. Why, one fellow even tried to make shoes out of banana peel. But the tanned peel was still too soft for anything but slippers.

Time flies like an arrow. Fruit flies like a banana.

The shortest unit of time useful in physics is the bananosecond. It is the time between slipping on a banana peel and landing on the sidewalk.

Depleted uranium atom

92 tired protons
145± sickly misshapen neutrons
92 slow-moving, erratic electrons.
No half-life. Or so they think.

Normal uranium atom

92 vigorous protons
145± neutrons of high character
92 well-behaved electrons
scintillating personality
half-life = 4.5×10^9 y

4

THE IDEAL SCIENTIFIC EQUIPMENT COMPANY[1]

Super April Sale!

Specials for physics teachers

If you've ever been embarrassed by demonstrations that didn't (or if you've ever occasionally faked results) then you need *ideal* equipment. No teacher should be without an adequate stock of *ideal* equipment. We *guarantee* the textbook results *every* time. It's easy to use, inexpensive, and it works!

Frictionless planes

Essential equipment for all studies of motion and mechanics. All of Newton's Laws can be correctly and easily demonstrated with the aid of an ISE frictionless plane (together with some or all of the other items shown in this catalog). Available in any size up to $3\,\text{m} \times 2\,\text{m}$! Just send us your specifications!

SALE PRICE . . .$16.95/m^2 (Cutting charges extra.)

[1] From a spoof by Kenneth Woolner, extended, with permission, by Donald Simanek and John Holden. The shorter version appeared in *Phys-13 News*.

Point particles

Another "must" for proper mechanics demonstrations! ISE point particles (integral numbers of grams), are available in all masses from 1g to 10 kg. Buy the 10 kg box and SAVE!

SALE PRICE . . .$11.65/10 kg

Hydrodynamical fluid

We are the sole supplier of this unique product (made by us under license granted by Bernoulli International). Completely free of viscosity and totally incompressible, it is otherwise indistinguishable from ordinary water.

SALE PRICE . . .$850/bbl.

γ-Ray microscopes

A modern realization of a classic! Heisenberg microscopes, superbly hand-crafted in the workshops of our European subsidiary, Gedanken Wissenshaft. Just a few left, so order today! UNCERTAINTY GUARANTEED!

SALE PRICE . . .$743.85 ea.

Inextensible string

A popular standby. No laboratory is complete without it! GUARANTEED to satisfy all textbook specifications for pulley problems. This string is completely inextensible, weightless, and perfectly flexible. Buy it by the 200 m reel.

SALE PRICE . . .$6.47/reel

Inertialess pulleys

Everybody said it couldn't be done! We are extremely proud of this new addition to our catalog, and we are certain that ISE inertialess pulleys will soon revolutionize the art of physics demonstration. Price proportional to square of radius. Use with frictionless bearings (below) for best results.

SALE PRICE (1 cm radius)...$7.69 each.

Frictionless bearings

Watch them go round and round forever!! These are the same bearings used in our perpetual-motion machine of the first kind.

SALE PRICE...$12.45/pr.

Rigid bodies

Have you ever had a demonstration fail due to non-rigidity of an essential part of your apparatus? This can be both frustrating and embarrassing, but it is a soluble problem! An ISE rigid body NEVER flexes! We can cut and machine this material to any desired specifications. Specify precise dimensions when requesting an estimate.

SALE...20% off regular prices.

Standard horses

These magnificent animals represent the culmination of seventy years of selective breeding experimentation. Each horse is GUARANTEED to do work at a rate of 746 W or your money back!

SALE PRICE...$1599.95 ea.

The horsepower of our standard horses is rigorously measured in our state-of-the-art testing laboratories.

Newtonian apples

Demonstrate a small part of physics history and complete your core curriculum! These apples can also serve as treats for your Standard Horse. Harvested from trees grown from scions of the original apple tree in Newton's backyard. Subject to the strictest standards of quality control, each box contains 4 dozen apples, individually wrapped.

SALE PRICE . . .$19.85/box

BONUS! **FREE**

With every order over $35.00 we will send you, ABSOLUTELY WITHOUT CHARGE, one cylinder (5 liters at 25 atm.) of IDEAL GAS. Demonstrating gas laws need never again give you heartburn!! Manufactured exclusively by IEC.

IDEAL SCIENTIFIC EQUIPMENT COMPANY

"If you didn't get it from us — it's not IDEAL"

Millennium Sale

FOR SERIOUS RESEARCHERS WHO WON'T SETTLE FOR LESS THAN IDEAL RESULTS. Our line of research apparatus is second to none, because it's **IDEAL!**

Perfect heat sinks

Our standard model heat sinks now have infinite capacity ($\pm 2\%$). If you have the time and the energy we invite you to test their capacity yourself.

$3000.00 each

New Improved Model Perfect Heat Sink is supplied in Formica with simulated chrome trim. $4000.00 each

Trivial solutions

Perfect for cleaning up those seemingly insoluble problems messing up your lab. $10/liter

Zeno's paradoxes

Frankly we don't know what uses these might have. We bought the entire inventory of a bankrupt manufacturer. They are cluttering our warehouse so we have priced them to move out fast. You might say these are budget paradoxes.

$1/pair (as is, no guarantees)

Carnot engines

These engines operate silently and never need lubrication. Guaranteed efficiency of $(1 - T_c/T_h)$. Can be run in reverse to cool the lab in summer. Use with our perfect heat sinks for best results.

$2000.00

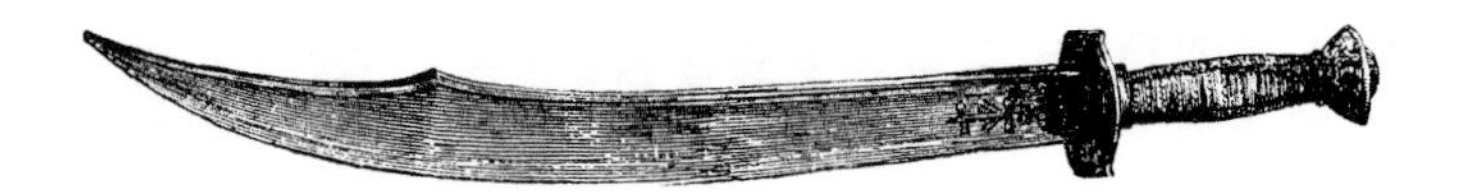

Occam's original razor, from the Ideal Equipment Company museum.

Occam's razor

The basic model introduced by Occam has been greatly improved by the Swiss craftsmen in our Hong Kong manufacturing facility. Blades of finest Swedish surgical steel (imported from Japan). Perfect for slicing through obfuscation, severing red tape, and cutting up a colleague's pet theory.

$75.00 ea. Extra blades $5/doz.

Occam's razor has many uses.

Black holes

A space-age concept finally brought down to earth. Every lab needs one, and one is quite sufficient, so we sell them singly only. Thousands of uses. A perfect way to dispose of your used Occam's razor blades, unwanted surplus equipment, OSHA inspectors and annoying lab visitors. Replaces administrator's paper-shredders—will even neatly dispose of the old shredder. Permanently dispatches into oblivion unwanted lab data which won't fit theory.

$999.99 each

Sorry, not available in colors, only basic black.

WARNING: The Surgeon-General and the FDA have not approved these for use in holistic medicine.

Sold only in our special shipping container which also serves as a safety shield:

$4,000,000.00

Perpetual-motion machine

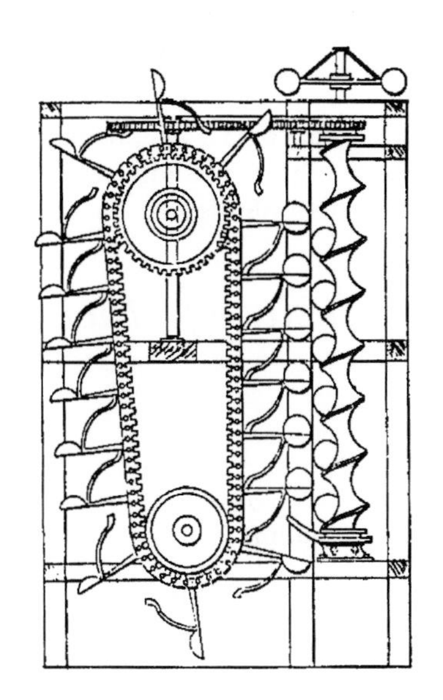

This precision machine will be a real conversation piece sitting on your desk or lab bench, turning silently and forever. Crafted of gleaming brass and chrome on a polished walnut base. Absolutely noiseless and vibration-free. While entropy increases all about you it is comforting to relax and watch this precision mechanism spin eternally. This is a perpetual-motion machine of the first kind, therefore it is in continual motion but produces no useful work. This might be a metaphor for much of what goes on in the modern workplace. Includes plastic dust-cover. Power cord not required.

$999.99

Quantum mechanic's tool kit

The essential tools of the trade, packed in an elegant attaché-style simulated leather case. Indispensable for tinkering up wave functions.

Geiger-Nuttall rule.

- Miniature wrench and screwdriver sets for effecting small perturbations (in quantized sizes).
- Crank for equations.
- Torque-wrenches in integral and half-integral sizes.
- Generous supply of assorted Hermitian operators.
- A large bottle of our best quality nuclear glue.
- Precision try-square for testing orthogonality of wave functions.
- Steel Geiger-Nuttall rule.

$1499.00

SPECIAL BONUS OFFER

All orders over $500 postmarked before April 30 will receive free of charge this new audio recording on two compact disks. Relive the history of physics in sound!

Music of the Spheres. Free

Includes:

Harmonice Mundi, by Johannes Kepler (B.V.D. No. 3782). The Angelic Choir of Vienna accompanied by Agnes Dee on the Celestial Harmonium.

Mysterium Cosmographicum by Johannes Kepler. Performed on period instruments by the Canonical Ensemble of Stuttgart.

The Harmonious Oscillator by Erwin Schrödinger. Realized on the Dirac Polyphoney Synthesizer using strictly orthogonal wavefunctions with relativistic modulation.

Perfect for background music in lab. Especially appropriate for planetarium shows.

Due to renewed interest in analog vinyl recordings, we also offer this as a 12 inch LP. Manufactured exclusively for us by Angelic Records. It achieves the remarkably long playing time of 75 minutes as a result of a revolutionary ($33\frac{1}{3}$ rpm $= \frac{5}{9}$ Hz) "overlapping grooves" cutting process.

This recording, in either version, may also be purchased outright for $40.00.

IDEAL SCIENTIFIC EQUIPMENT COMPANY
"If you didn't get it from us — it's not IDEAL"

5
CARTOONS

"So what if all the journals rejected it? We can still use it on the next freshman physics exam."

A cartoon similar to this was posted on a classroom bulletin board without caption, with an invitation to physics students to supply their own caption. Here are some of the best ones:

"Now let's see the error analysis for that."

"See, I told you the proof was obvious."

"Now that you've stated the problem, what's the solution?"
"I'm afraid I see an error back there in your second step."
"Wouldn't you know, when we finally got the answer, it would be something that simple!"
"Yes, that's much clearer."
"Amazing! I was sure that two tablespoonsful would suffice."
"Does this apply always, sometimes, or never?"
"That reminds me, my wife wants me to pick up some alphabet soup for the kids on the way home."

Spider physics

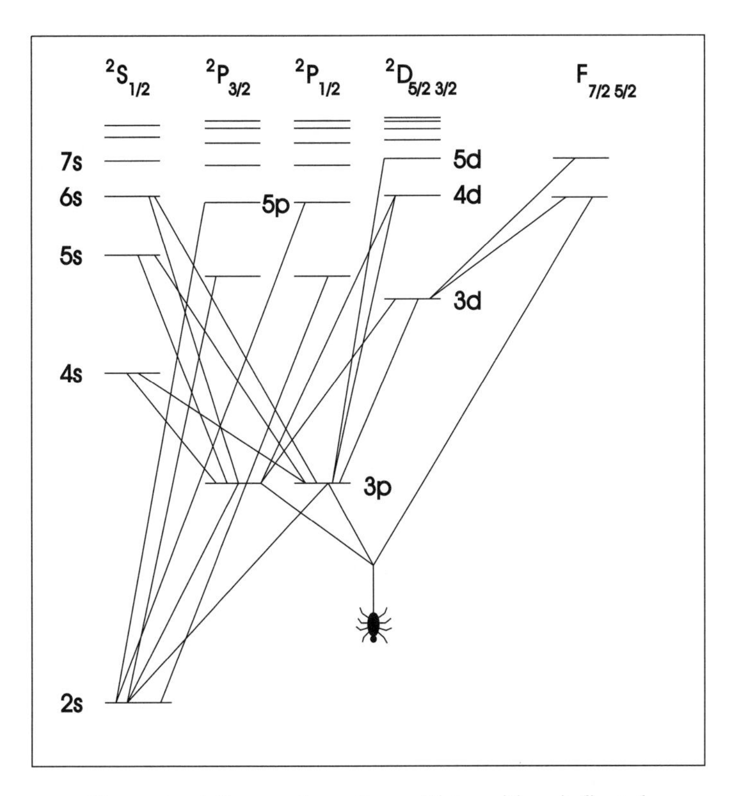

Energy level diagram for sodium with transitions indicated.

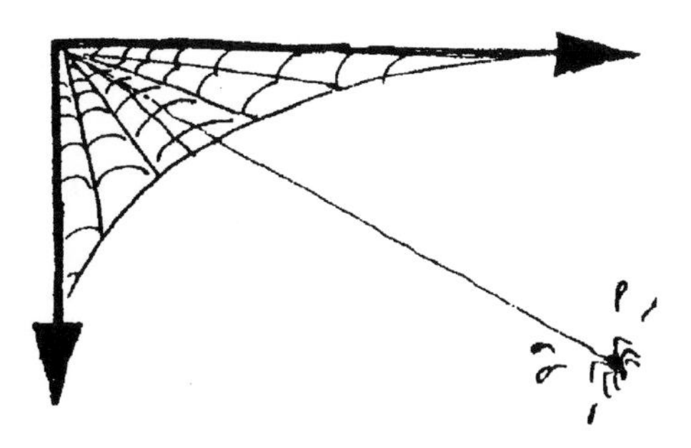

May the NET force be with you.

Dr. Watson: "Surely, Mr. Holmes, you know the function of laxatives."
Mr. Holmes: "Alimentary, my dear Watson, alimentary."

Department of UNCLEAR Physics

"I don't care if the force IS with you, so long as the NET force is with ME!"

6

GEOLOGY

Geology made simple

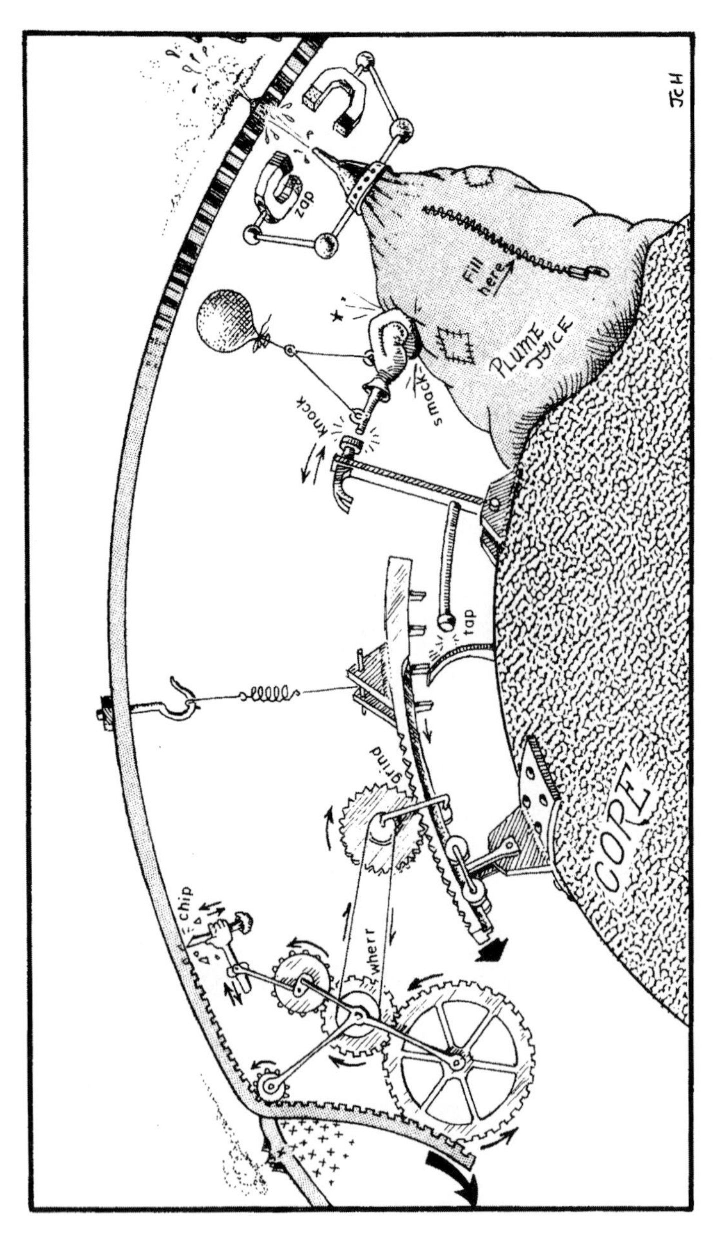

From EOS, Trans. Amer. Geoph. Union 58, 7 (July) 1977.

To date, tens of thousands of scientific man-hours have failed to discover why the continents allegedly drift and why plate tectonics might exist. Here we see the clockworks of the earth's interior and how plumes and sea-floor spreading (far right) are synchronized with subduction (far left). Now we know how the earth *really* works.

With the advent of plate tectonics in the 1960s, it became apparent that the earth is tectonically "alive"; its crustal carapace continuously renewing along the 40,000 km long rift zone coinciding with the global mid-ocean ridges. The process of crustal rejuvenation is stylized here as *terrorostracodus tectonica* Simanek and Holden, sp. nov. sheds its carapace, breaking apart in tectonically acceptable areas.

Fake Tectonics and Continental Drip[1]

by JOHN C. HOLDEN

With the advent of plate tectonics theory in the late 1960s (Morgan, 1968) geologic thought was revolutionized; continental drift gained popularity (at last) and, as everybody has been saying: "the textbooks must be rewritten." Unfortunately, no one to date has done it right.

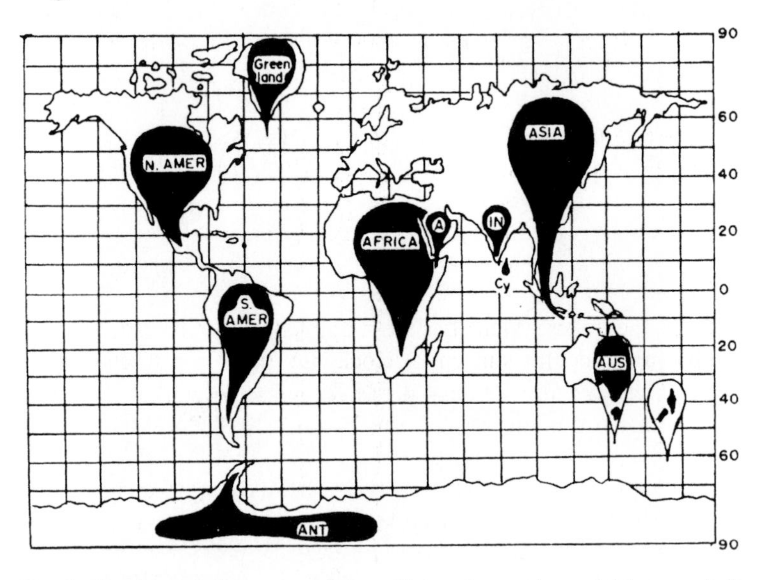

Fig. 1. Continental shapes reveal the evidence for continental drip, especially when this figure is viewed upside-down.

There is a very important dimension missing in the new theory, one so obvious that it is little wonder that it has escaped the attention of geologists and geophysicists. I refer to the shapes of the earth's continents and their bearing to past drift motions. Conventional plate tectonics states that continents drift, relative one to the other, because they are situated on the backs of crustal

[1] Reproduced by permission from the *Journal of Irreproducible Results*, **22**, 2 (July) 1976.

plates, which move in response to sea-floor spreading away from mid-ocean spreading ridges into subduction zones where the plate disappears into the earth's interior. The whole process is similar to the activities of many conveyor belts joined on the earth's surface in constant, though slow, motion.

However, the outlines of all continents are tear-drop shaped with the pointed ends trending south indicating the directions from which they have come. This startling concept is called "continental drip" and is the essential concept of "fake tectonics," the acronym for Final Answers for the Knowledge of Earth tectonics. Actually, fake tectonics can be traced back to the early historical development of continental drift when Wegener (1929) referred to the continent's flight away from the poles, or "polarfluchtkraft." A step in the right direction was made by Dietz and Holden (1970) when they recognized that there was only a "sudpolarfluchtkraft" (south polar fleeing force). The whole truth can now be recognized. There are "sudpolarfluchttrofeln" (south polar fleeing drips).

The concept has great heuristic value. For instance, because the continents appear to flow north we can finally recognize that our place in the solar system has been topsy-turvy—that the planets revolve around the sun in a clockwise (not counterclockwise) manner and that the south pole should be on the top of the globe and the north pole on the bottom because, as we all know, drips should always flow down.

Figure 2, a map of the world from a modified cylindrical equally spaced-out projection (Deetz and Adams, 1945) shows that all continents are fundamentally drip-shaped. Some show a multiple drip aspect; the subdrip of Arabia being a good example (A). The open drip symbol of New Zealand is based on its submerged continental outline and must be considered a drowned drip. The anti-drip of Ceylon (Cy) is interesting and is included as it is the exception that proves the rule. The anti-drip of Antarctica is due to the fact that Antarctica being situated at the top of the globe has not yet made up its mind which way to flow hence that continent has drifted round and round but never down.

Considering that the universal landmass of Pangaea broke up some 200 million years ago, and that the continents are now better than halfway dispersed from the original nucleus at the south pole, the time of their re-uniting at the other end of the world can easily be predicted. Based on computer analyses of the northward

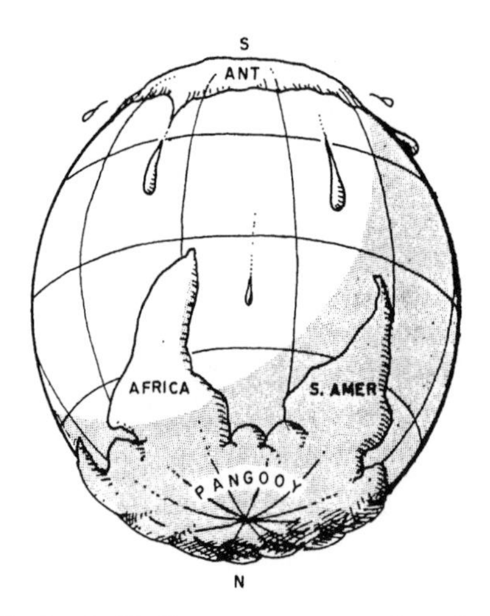

Fig. 2. The inevitable Pangooy outcome of continental drip.

vector sums for continents now in motion, this event will occur on Tuesday morning at 9:00 a.m. 1,786,379 A.D. The resulting new universal landmass (excluding Antarctica) is called "Pangooy" and the event is termed "continental splash"; the final coalescence of continental drips. Insofar as Antarctica is perfectly balanced on top of the globe, it is highly unlikely that this continent will ever join the others on their trek down under.

References

Deetz, C. H., and O. S. Adams, 1945. Elements of map projection. *U.S. Coast and Geodetic Surv.*, Spec. Publ. 68, 226 pp.

Dietz, R. S. and J. C. Holden, 1970. Reconstruction of Pangaea: breakup and dispersion of continents, Permian to present. *J. Geophy. Res.*, vol. 75, pp. 4939–4956.

Morgan, W. J., 1968. Rises, trenches, great faults, and crustal blocks. *J. Geophy. Res.*, vol. 73, pp. 1959–1982.

Wegener, A., 1929. *The Origin of Continents and Oceans* (transl. by J. Biram), 4th ed., 246 pp., Dover, New York.

Scraps from the talis slope

If rocks aren't romantic, then why is there carbon dating?

Geologists have their faults, of course. But actually they are misunderstood. The more they try to be gneiss, the more they are taken for granite.

Never lend a geologist money. They consider a million years ago to be recent.

A geologist is the only person who can talk to a woman and use the words "dike", "thrust", "bed", "orogeny", "cleavage", and "subduction" in the same sentence without facing a civil suit.

Geologists are amazing. They know hundreds of words for different sorts of dirt and hundreds of words for things it does when left alone for a few million years.

Geology field trip.

7

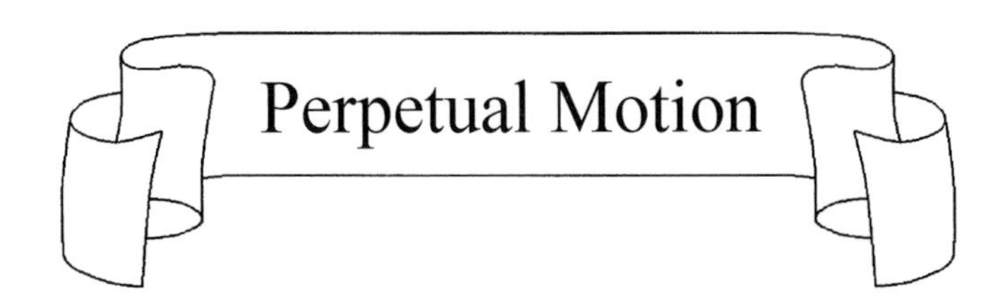

The Perpetual Search for Perpetual Motion

When people say, "It'll never work," they aren't always wrong. Most new ideas turn out to be wrong and are swept under the rug of history. Even wrong ideas can be useful, for they give us valuable information about what *doesn't* work, thus narrowing the field of things which *might* work. But some people never give up on a discredited idea. Such is the case with the perennial search for perpetual motion.

A perpetual motion machine is easy to define, difficult to make. The classical kind is a machine that continues its motion forever, without input of energy. Most inventors want more—a machine that produces more energy than it takes in, in violation of the first law of thermodynamics. Such a machine would run forever *and* produce excess energy to run other things. Part of its output could even be used to drive the input, so it requires no energy source at all. Such a machine could make the energy crisis disappear.

Much human ingenuity has been expended in a long succession of ingenious failed attempts to build such machines. Such is the tenacity of the human spirit that men still persist, undaunted by past failures and undeterred by the laws of physics. Over the centuries clever inventors devised a bewildering variety of schemes to thwart nature's perversity. Lesser intellects repeated these mistakes, stubbornly trying to re-invent the square wheel.

French architect Villard de Honnecort proposed such a machine in the year 1245. He described an overbalanced wheel with hinged hammers equally spaced around its rim. As the wheel turned each hammer flipped over to a new position as it moved over the top of the wheel. This mass transfer (or perhaps the impulse

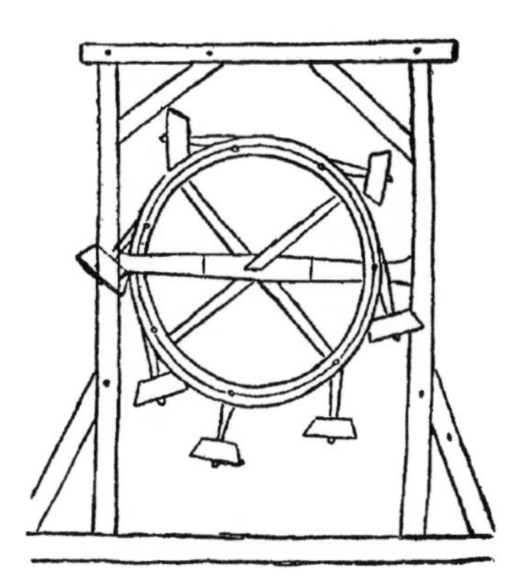

Honnecourt's overbalanced wheel, 1245.

due to its rapid motion) was supposed to give a boost to maintain the wheel's motion. Honnecort claimed his device would be useful for sawing wood and raising weights.

The diagram below suggests that the hammers on the right side of the axle have greater lever arms, so there must be a net clockwise torque. Therefore, the inventor claimed this would move clockwise.

All of these overbalanced wheel ideas can be shown to be faulty because they incorrectly represent the physics. If this wheel really were moving clockwise, the inertia of each ball would make it stay at a larger radius as it moves over the top of the wheel. A snapshot of it in motion wouldn't look like this picture. The static picture hides this fact.

Something in the perpetual motion idea tantalizes restless minds, who say, "Yes, previous attempts have failed. But what if there's some principle of physics as yet undiscovered which we might reveal if we just tweak the design a bit." So, through the centuries, they did fiddle with the details in a fantastic variety of ways until the proposed machines became fiendishly complex. Still they didn't work.

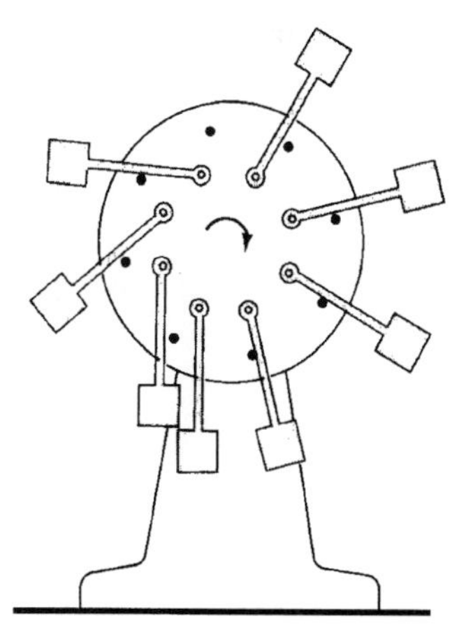

Honnecort's overbalanced wheel.

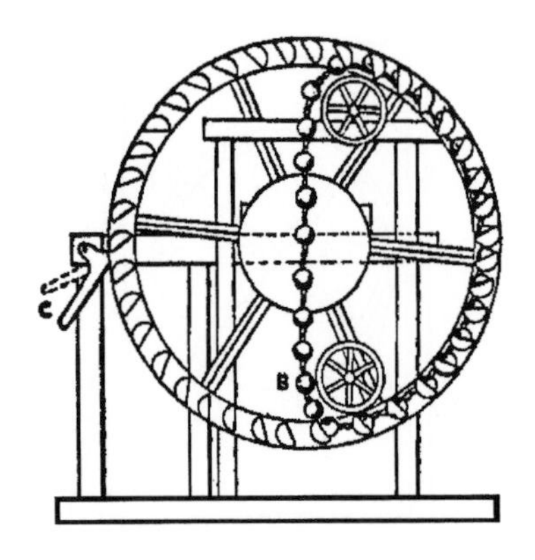

Perpetual motion machine with safety brake.

Perpetual motion machine inventors are nothing if not optimistic. Some were so certain that their designs would work that they included a brake to prevent the wheel from rotating fast enough to tear itself apart. They needn't have bothered.

Some design drawings included an arrow to indicate which direction the wheel was supposed to turn. This is very helpful. Without it we wouldn't have the slightest idea which direction the wheel should turn.

Most designs are derivative of earlier failed proposals. The inventors were not very imaginative. There's no reason why a device should have to have moving parts in order to attain efficiencies greater than one. Perhaps the very name "perpetual *motion*" has

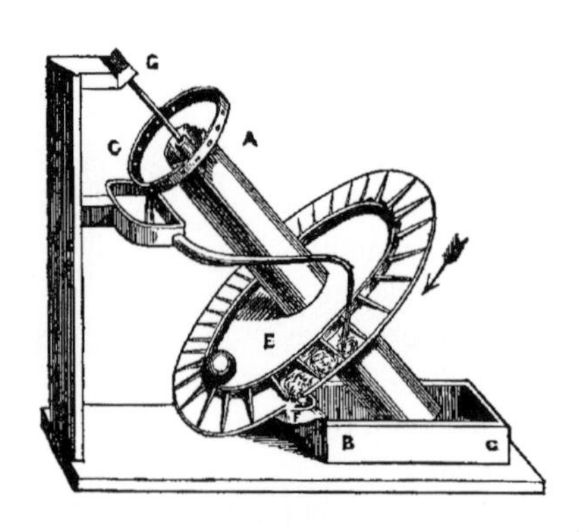

A helpful arrow.

limited their thinking to devices which move. Curiously we see hardly any proposals of a purely electronic device, sort of a super-transformer, which puts out more power than it takes in.

The perpetual search for perpetual motion never stops. Each year, some new device is reported with sensational newspaper head-lines proclaiming that it achieves over-unity efficiencies in energy conversion. One headline announces a magnetic machine supposed to have 135% "apparent efficiency." One can be sure it is only "apparent."

Many of the inventors of perpetual motion machines are sincere but manage to delude themselves. Others are outright frauds hoping to delude others. Then there are those who use other methods of deception, for humorous result, as in the following examples.

Mathematical Perpetual Motion

Physicist George Gamow devised this "mathematical perpetual motion machine."[1] Since 9 is greater than 6, and rotation of the wheel turns the 9s into 6s (and vice versa) the wheel should turn in the direction shown.

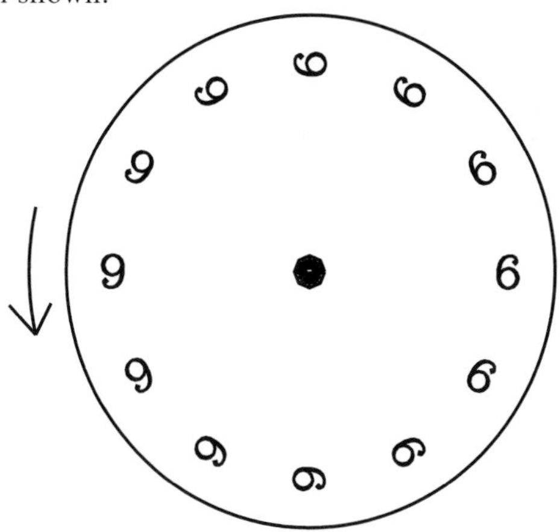

[1] Though this has been attributed to Gamow, mathematical jokes of this kind have been going around for many years. The original genius who devised this revolutionary idea may never be known.

Before you dismiss this ingenious idea out of hand, consider the fact that the principle of this machine also works just as well if one chooses appropriate numbers expressed as Roman numerals.

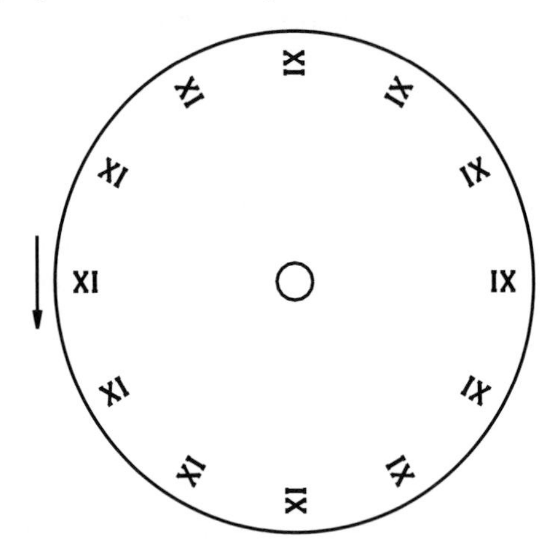

There's a subtle principle these wheels have in common. Martin Gardner used the same principle in the up–down engine shown below, which surely works just as well as the others.[2]

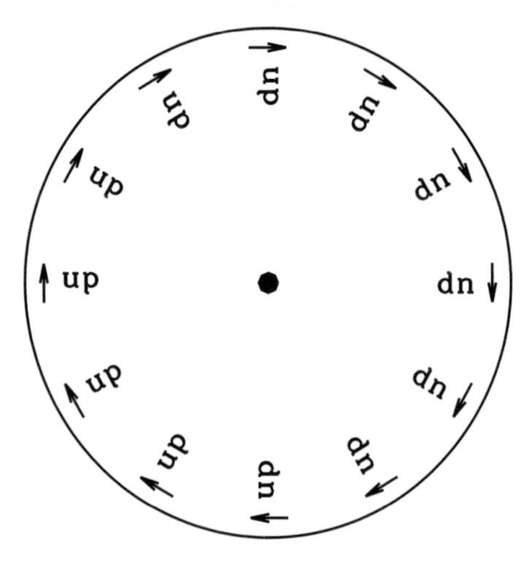

[2] Gardner, Martin, in "Mathematical Games," *Scientific American*, Feb. 1972. Also see his book *The Magic Numbers of Dr. Matrix*. Perceptive readers will find other applications of Gardner's principle elsewhere in this book.

These are based on previously neglected principles: the principles of mathematical and artistic deception applied to the endless quest for perpetual motion. We have extended this sort of deception to the time-honored overbalanced wheel, (self-impelling wheel). Many early perpetual motion machines used compartments with rolling balls inside a structure which, the inventor thought, caused more balls to be always on one side of the axle so that side of the wheel would be continually heavier. That didn't work. But if one warps the compartments (or space itself), it just might work.

Perpetual motion machine with balls in compartments.

We have applied this idea to a self-impelling wheel with heavy balls moving freely in closed compartments. The outer panels of these compartments have been removed in this drawing to show the balls inside. We don't have a working model yet, because of difficulties encountered while engineering those panels and affixing them to the main frame of the wheel. It is not easy to design a proper axle, so we propose that the main drive wheel roll on two supporting wheels below. These could be attached to a drive shaft or pulley, using standard engineering methods to transfer the power for useful purposes.

The wheels which support the drive wheel could also serve as drive wheels on a vehicle. The prototype concept drawing does not show any method for speed control, or for shifting into

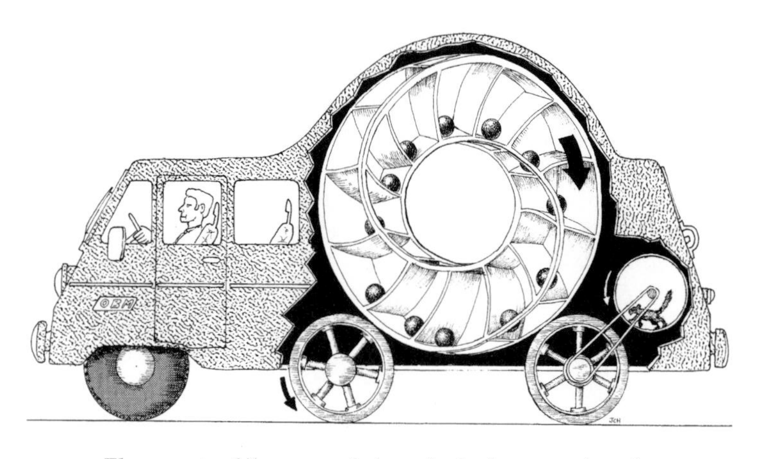

The perpetmobile, an overbalanced-wheel perpetual motion vehicle with balls in compartments.

reverse. It does show the starter mechanism in back, which all such perpetual-motion wheels require. This one is of the classic "squirrel-cage" design, complete with squirrel. (This is actually an inside joke perpetrated by someone on the design team who thought the whole idea was a bit squirrely.)

One of our young hot-shot engineers, fresh out of school, suggests we use a movable magnet inside the wheel, acting on the steel balls. Then when the driver wants to change speed, or go in reverse, one need only shift the position of the magnet. The details of the implementation of this idea cannot be revealed at this time.

Research continues to find a way to insulate the passenger compartment from the continual noise of the clattering balls.

Our development team is confident these minor engineering details will be solved, thus creating the first perfectly non-polluting means of transportation. At last our dependence on fossil fuels can be ended.

Example from the real world

When parody becomes indistinguishable from reality, we have reason to be concerned.

DES

Attempting to make fun of pseudoscience is a difficult task, for one must compete with pseudoscientists producing funnier material while being absolutely serious. Consider this unsolicited e-mail message:

To whom it may concern:

I have developed a renewable energy source. This method generates one to hundreds of thousands and even millions of pounds of pure thrust. There is no fossil fuels or nuclear power required to generate this thrust. There also is no waste. If one was to choose to take waste from this method the waste would be pure.

This thrust is generated by a type of perpetual motion that I have developed. This motion generates the thrust that could turn anything it is connected to.

I am concerned that the whole theory might be muffled by companies/corporations of great power. And that is not my wish.

Concerned Inventor,

(Name withheld to protect this lone genius from being muffled by those ruthless companies/corporations of great power.)

"It may be perpetual motion, but it will take forever to test it."

8
BIOLOGY

When a problem gets too complicated for the physicists, they hand it to the chemists.

When a problem gets too complicated for the chemists, it is handed over to the biologists.

And when biologists think it is too complicated, they give the problem to the sociologists.

The bear facts

In "Bailliere's Comprehensive Veterinary Dictionary" by D. C. Blood and Virginia Studdert, the following entry appears on page 133, right between Brunner's glands and brush border:

> Brunus Edwardii: *the urban, companion animal bear, much admired for its low food requirements and excellent house training, a high emotional output and complete freedom from disease. Called also* Ursus Theodorus *(USA) and Pooh, Paddington or Brideshead bear (UK).*

Symbiotic attachment

In a neurobiology lecture the professor mentioned that much of the data he had shown were culled from studies of leeches. He said, "Now, a lot of you may think leeches are nasty creatures. The people working with these creatures are quite fond of them, however. It is also reported that sometimes the leeches even become attached to the researchers."

That's life

Life is a sexually transmitted disease.
Life is anything that dies when you stomp it!

All in the eye of the beholder

A biology professor was addressing his class, wanting to see if they'd read the assigned text. He asked Miss Smith to stand. She does.

Professor: Miss Smith, what part of the human body increases in size ten times when excited?

Miss Smith blushes and hesitates and giggles.

Professor: Miss Smith, please sit down. Miss Jones, please stand and tell me if you know what part of the human body increases in size ten times when excited.

Miss Jones: Yes, Professor. It's the pupil of the eye.

Professor: Very good. Thank you Miss Jones, you may sit down. Miss Smith, will you please stand again. I have three things to say to you.

1. You have not done your homework.
2. You have a very dirty mind.
3. You're in for a big disappointment.

Brain power

After many years of failure, science finally perfected a process for replacing an individual's brain with brain tissue from other individuals. At last it was possible for a person to replace inferior brain matter for some of superior quality.

A college student went to the campus brain-transplant clinic to inquire of the cost of an upgrade. He noticed a clearance sale sign: "Special: Politician's brains, $1 per ounce, no refunds, no guarantees."

The salesman explained, "We've got a plentiful supply of them, and they've been hardly used at all, but the quality is uncertain."

"Well, I don't want any of those, even at that price," the student said. "I want something with great intellectual power and ability."

The salesman got out his price book, and said, "We can give you the brains of a college instructor for $10 an ounce, or a full professor for only $100 an ounce."

"What if price were no object?" the customer asked.

"Then you might choose deans' brains at $100,000 per ounce, or college presidents' brains at $1,000,000 per ounce."

The student was astounded. "Why such a big difference between a professor and a college president? There can't be *that* much difference in quality."

"The reason is easy to see," the salesman replied. "Have you any idea how many college presidents it takes to get one ounce?"

The Reconstruction of "NESSIE": The Loch Ness Monster Resolved[1]

by JOHN C. HOLDEN

Introduction

In recent years there has become available to science an increasing amount of important data concerning the existence of a hitherto undescribed animal residing in the body of fresh water known as Loch Ness in northern Scotland. It is now possible to reconstruct a close facsimile of this enigmatic organism based on three types of data: (a) direct factual information obtained by scientific inquiries on the subject, (b) rumors and hearsay about such a creature and the characteristics ascribed to it by local residents, and (c) inductive logic consistent with the geological, paleontological, and biological requirements of the situation. Within the limitations of the above constraints, it is now possible to place the species commonly known as "Nessie" into the scheme of contemporary taxonomy and once and for all establish the scientific reality of the Loch Ness monster.

---

[1] Reproduced by permission from *The Journal of Irreproducible Results*, **20**, 4 (July 1974).

Fig. 1. Sketch of the Loch Ness monster according to information supplied from an eye witness. Due to the trauma of the experience many details of this reconstruction may be in error, though gross morphology is considered valid.

According to a recent report by the Loch Ness Investigation Bureau [1] sonar signals were recorded from hydrophone experiments in Loch Ness. These noises, from as deep as 300 feet in the Loch, consist of sonic vibrations of variable intensity and frequency. Some 25 miles of recording tapes still await interpretation. Scientific activity on the Loch Ness monster is of fairly recent origin. The legend of the monster goes back much farther, however. In general, first hand observers have described Nessie as a slithery or undulating object breaking water. These data are usually gained at night, especially on foggy or other evenings of low visibility attesting to the animal's keen shyness to being too closely scrutinized. One source [2] has described his personal encounter to the author which is here summarized in Fig. 1. There are undoubtedly some errors in the sketch as the witness was highly excited during the interview. It is interesting to note that during the Second World War the German High Command had sufficient confidence in the reality of the monster to actually drop bombs in Loch Ness with the intent of destroying the creature and, thereby, damaging British morale [3].

In this paper, most of the emphasis is placed on the inductive logical parameter mentioned above. With these data a realistic interpretation concerning the nature of Nessie can be made in keeping with the observations and being at the same time consistent with taxonomic and geologic theory.

Two animal types are good candidates for the alleged monster: one, cetaceans (porpoises and whales) and two, plesiosaurs, a thought-to-be-extinct group of marine reptiles. The cetaceans can attain "monstrous" size and are known to have developed complex sonar systems. We must rule them out, unfortunately, since according to Marshall, the noises in Loch Ness are unique. The plesiosaurs on the other hand, are a fruitful avenue of investigation. First, they show no paleontological evidence of ever having developed sonar signals. Therefore, should they have done so it would be unique. They also have the additional advantage of being more naturally monstrous looking than the friendly cetaceans. The plesiosaurids were large reptiles and, though not related to dinosaurs, were of comparable size and must have been the dragons of Jurassic–Cretaceous seas.

Classification and systematic description of the species

Taking the basic plesiosaurid shape, which has been aptly described as "threading a snake through the body of a turtle" [4] as a guide, it is possible to make the appropriate changes and reconstruct what Nessie probably looked like. The important features of the reconstruction include: (a) a well developed sonar organ, (b) loss of eyes, (c) elongation and narrowing of the snout with the forward migration of the nostrils enabling the animal to breathe inconspicuously without exposing very much of its body, and finally, (d) basic modifications of the body morphology from the muscular, trim, streamlined ocean going model to the lethargic, flabby, freshwater type. The following formal description is a convention required by the International Congress of Zoological Nomenclature.

Phylum	CHORDATA
Class	REPTILIA
Order	SAUROPTERYGIA
Suborder	PLESIOSAURIA
Superfamily	PLESIOSAURIDEA
Family	PLESIOSONARIDAE new fam.
Genus	PLESIOPHONUS new gen.
Species	HARMONICUS new sp.

Plesiophonus harmonicus new gen., new species Figs. 1, 2, 3, 4b.

Description. Body size: large. Appendages: the *flipperae* and *flapperae* are typically plesiosaur-like; *neckus elongatus* long, sinuous, scary; *tailus terminus* with horizontal stabilizers. *Humpae* of various types new to science (see especially Fig. 3): A *humpus undulatus* in the shoulder region housing oil-filled sonar resonator

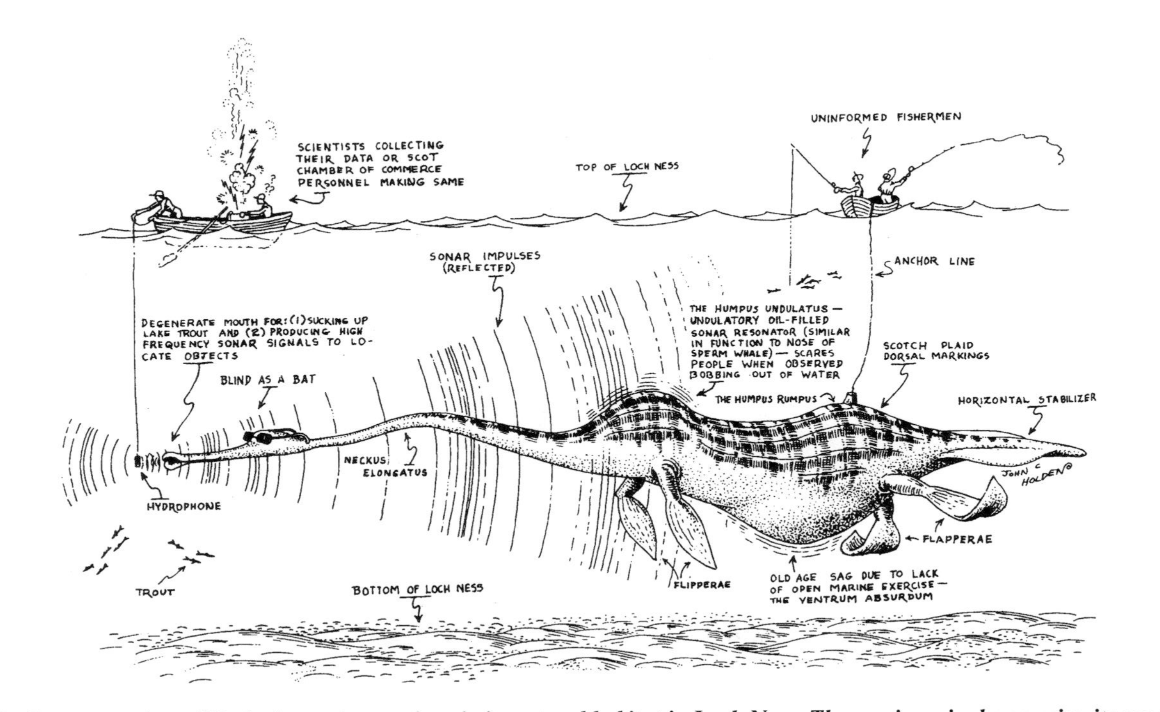

Fig. 2. Reconstruction of Plesiophonus harmonicus *in its natural habitat in Loch* Ness. *The specimen is shown using its sonar for locating a hydrophone. Other organisms indigenous to the area are also shown, for scale.*

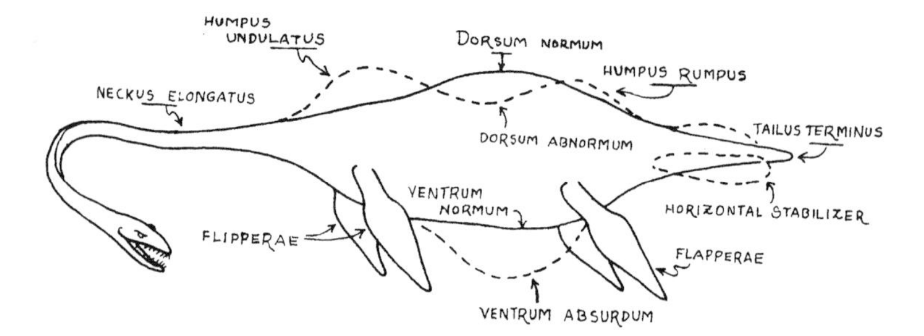

Fig. 3. The generalized body of **Plesiophonus harmonicus** *compared to the basic plesiosaurid type. The typical plesiosaurid body is shown in solid line; that of* **Plesiophonus** *in dashed line. The single* **dorsum normum** *of the plesiosaurid gives way to two* **humpae**, *an anterior* **humpus undulatus**, *and a posterior* **humpus rumpus.** *Also, the* **ventrum normum** *typical of the Jurassic-Cretaceous reptiles atrophies into a* **ventrum absurdum** *in the Loch Ness monster.*

sensitive to self emitted sonar frequency signals. This organ is extremely repulsive and strikes fear into those observing it bobbing out of the water on dark nights. Dorsum shifted posteriorly to form the *humpus rumpus* counteracting the anteriorly situated mass of the *humpus undulatus* and maintaining the animal's center of gravity. Prominent *ventrum absurdum* holds vital organs, developed as a result of the species' languishing listless loch life lacking proper ocean marine exercise. Head region (see especially Fig. 4b); highly specialized teeth and eyes absent. Anterior cranial parts elongated, forming snout-like mouth for sucking up lake trout and emitting sonar signals. Skull bulbous to accommodate increased I.Q. necessary for (a) acoustical higher mathematics, (b) keeping away from people, and (c) responding to ever more sophisticated electronic apparatus placed into Loch Ness by curious scientists. Markings: dorsum covered with Scottish plaid color pattern, naturally.

Natural history of the Loch Ness Monster

Loch Ness is the body of water occupying a very ancient and fundamental geological rupture known as the Great Glen Fault. This fault is a tectonic juncture within the crustal Paleozoic and older rocks comprising the basement of northern Scotland. The fault is very

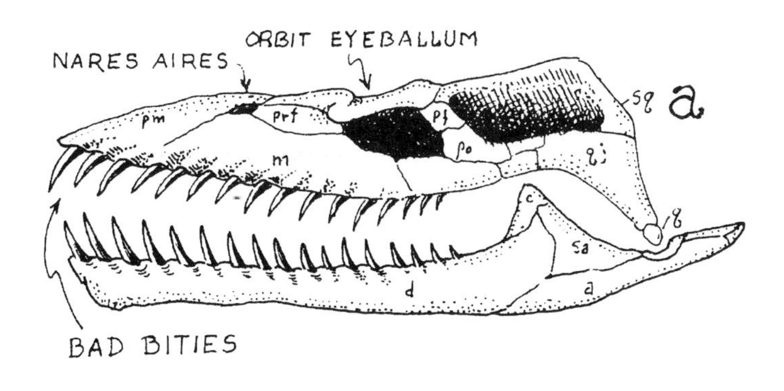

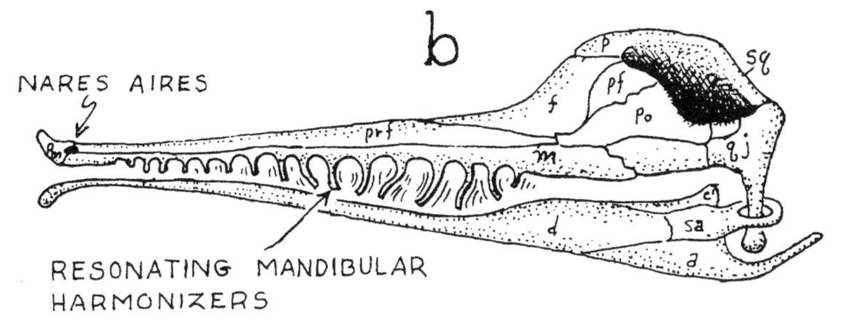

Fig. 4. Contrasting skulls of the extinct plesiosaurs (4a above) and closely related* Plesiophonus *(4b below) from Loch Ness. In the latter the teeth are no longer needed and have disappeared through disuse. Similarly, the* orbit eyeballum *is gone with the concomitant enlargements of the* postfrontal *(pf),* postorbital *(po), and* frontal *(f) skull segments. The* maxilla *(m) is modified into a series of tuning forks that enable a wide range of unique sounds utilized in sonar emission. The* prefrontal *(prf),* maxilla *and* dentary *(d) are elongated to form a prominent snout.

ancient as so must be the loch. In fact, Wilson has suggested that it is contiguous under pre-continental drift reconstruction with the Cabot Fault in the Bay of Fundy, North America [5]. This makes the loch greater than 220 m.y. old. The plesiosaurid great reptiles existed after the basin was already in existence in the Jurassic and Cretaceous (180 to 65 m.y.b.p.) so probably inhabited the narrow seaway that extended into the Great Glen Fault. At the end of the Cretaceous the basin was closed to the sea and gradually became freshwater. Most of the entrapped marine fauna became extinct except for some plesiosaurs which adapted to the new freshwater conditions and decided to remain there. Today they titillate the curiosity of visiting scientists, scare the local Scots, and bolster the national economy with visitors from all over the world waiting for a glimpse of something new.

In order to adjust to the new freshwater conditions the entrapped reptiles had to learn to live in the murk of a freshwater body which lacks, by definition, the sodium cations so helpful in flocculating clays and other fine particulate matter. Therefore, early in the monster's life history, Nessie evolved a highly sophisticated sonar organ. As this happened the eyes degenerated by disuse. Competition dropped off markedly in the new environment as *Plesiophonus* found itself there alone. This lack of exercise produced phylogenetic muscular atrophy and the *ventrum absurdum* and *humpus rumpus* developed (not unlike in other vertebrates, e.g. hominids, under easy physical conditions). In the sleepy refugium of Loch Ness the species did not become extinct as did his cousins in the harsher ocean environments.

For more than 65 m.y., small changes withstanding, the species remained in homeostatic equilibrium with its environment. With the appearance of man on the scene it then underwent evolutionary saltation in response to intolerant and meddling hominid behavior. Due to specialized prejudicial activities of early man along the banks of Loch Ness, Nessie was induced to evolve a coloration pattern mimicking the traditional Scottish plaid. Undoubtedly, nonconformists were selected against violently. During this time the *nares aires* migrated anteriorly along and to the end of the elongate snout thereby enabling the species to always keep most of its head submerged.

It remains only a matter of time until one of the specimens of Nessie is captured. Motivation is now supplied not only by a desire for the scientific knowledge to be gained by a face to face encounter but also by a monetary reward. $1000 (U.S.) has recently been offered for the capture of the "monster" [6].

Literature cited

1. Marshall, N. B. (In Nichol, D. M.), 1970. "None can say Nessie has lochjaw". *Miami Herald*, Fri. Dec. 4:3-G.
2. Prof. Samuel P. Welles, 1971 (personal communication).

3. Dr. Robert S. Dietz, 1970 (personal communication).
4. Romer, A. S., 1968. *The Procession of Life*. World Publ. Co., New York, 195.
5. Wilson, J. T., 1962. Cabot Fault, an Appalachian equivalent of the San Andreas and Great Glen Faults and some implications for continental displacements: Nature, 195: 135–138.
6. Sneigr, D., 1970. "Miamian offering $1,000 for best 'monster'". *Miami News*, Wed., Dec. 9:7-G.

Always consider the obvious

A group of goose biologists were meeting to brainstorm about the migration behavior of Canada geese. They were particularly interested in applying for a $100,000 Federal grant to investigate the "V" formation of goose flight. It had been observed that one side of the "V" is always longer than the other side. This group would put together a research proposal to apply for the $100,000 grant and hopefully discover why this happens.

To start off the discussion, the Consulting Firm Biologist stood up and said in typical consultant fashion, "I say we ask for $200,000, and attempt to model the wind drag coefficients. We can have our geologists record and map the ground topography and then our staff meteorologists can predict potential updraft currents. Our internal CAD department can then produce 3-D drawings of the predicted wing tip vortices. Then, after several years of study, our in-house publications department could produce a nice thick report full of colorful charts and graphs."

The Senior Research Biologist, a professor at the local university, cleared his throat and responded, "No, no!, That's not it at all. We only need $150,000. We can train a group of domesticated geese to fly in formations of equal length and then compare their relative fitness to wild geese. We can publish the results in the *Journal of Wildlife Management*.

The field biologist rose and began walking toward the door. "Where are you going?" the group asked. "I'm leaving" he replied, "I've heard enough. No one has to give me $100,000 to find out that the reason one side of the 'V' is longer is simply because there are more damn geese on that side!"

☞ A mathematician might argue as follows. For there to be equal arms of the V, there must be an odd number of geese, since one is at the apex of the V. The chance of there being an odd number, rather than an even number, is very likely $\frac{1}{2}$, or 50%. Since the two arms of the V may be equal or unequal, this reduces the probability still more. In formations, one observes perhaps one to two dozen geese. Suppose there are 25. If the distribution of geese in the arms is unbiased, then there are 12 possible ways the geese can distribute between the arms, and only one of these has equal numbers in both arms. So the probability of the V having equal arms is $(\frac{1}{2})(\frac{1}{12}) = \frac{1}{24}$, or about 4% of the time. One might suspect that the probability is a bit higher than that, because the aerodynamic advantage may be smaller when the arms are very different in length. Still, we'd be surprised to see equal arms more than, say 5% of the time.

The frog who would be prince

A frog telephoned a psychic hotline and was told, "You are going to meet a beautiful young girl who will want to know everything about you."

"Great," says the frog. "Will I meet her at a party?"

"No," said the psychic. "Next year—in a biology class."

Field work

Policeman: What are you doing here?
Biologist: Just looking for flora and fauna.
Policeman: Move along, or I'll run you in along with your girl friends.

Could we even call it fertile research?

Something tells me that no matter how good a series of experiments on reproductive technologies, the paper written about the work will never be described as seminal.

Scraps from the dissecting tray

Support bacteria—it's the only culture some people have!

Thesaurus: ancient reptile with an excellent vocabulary.

9
ENVIRONMENTAL SCIENCES

Mankind has been despoiling the environment from earliest times, when Neanderthal man began drawing graffiti on cave walls.

Raping the landscape.

Later, as technology developed, the environment could be altered on a much larger scale.

Autoecology (From a limited edition print by John C. Holden, 1970.)

Another chance for evolution to get it right.

The population bomb. "Unplanned Parenthood" from R. S. Dietz and J. C. Holden **Creation/Evolution Satiricon.** *The Bookmaker, 1987, p. 124.*

If we continue polluting the environment with the wastes of our so-called civilization, we may hasten evolution in ways we can't anticipate.

☞ An oft-quoted cynical observation is "Mankind is a crawling disease on the face of the earth." From a purely physical point of view this is right on target. Considering the size of a bacteria relative to man, and man relative to the earth, the sizes of bacteria/man and man/earth are about in proportion. Also, since there are about 10^{11} microbes on a person's skin, the population/area ratio for microbes on man is about the same as the population/area ratio of man on the land area of the earth.

THE FOUR ELEMENTS

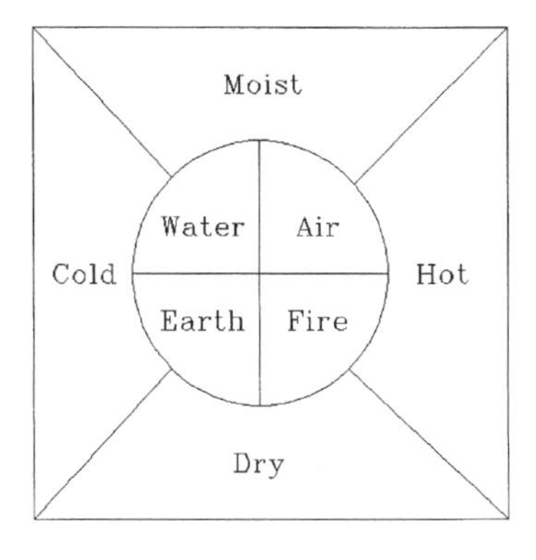

The four elements and their essences, as understood by Aristotle.

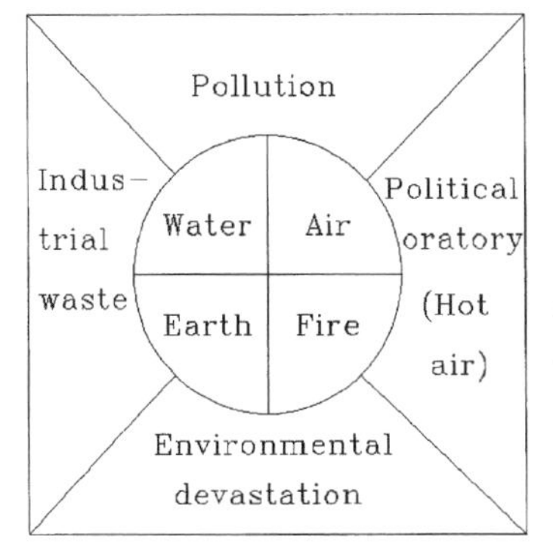

The modern version of the four elements.

Ban Dihydrogen Monoxide!

The Invisible Killer!

DAN GALVIN[1]

Dihydrogen monoxide is colorless, odorless, tasteless, and *kills* uncounted thousands of people every year. Most of these deaths are caused by accidental inhalation of DHMO, but the dangers of dihydrogen monoxide do not end there. Prolonged exposure to its solid form causes severe tissue damage. Symptoms of DHMO ingestion can include excessive sweating and urination, and possibly a bloated feeling, nausea, vomiting and body electrolyte imbalance. For those who have become dependent, DHMO withdrawal means certain death.

Dihydrogen monoxide

- is also known as hydroxyl acid, and is the major component of acid rain.
- contributes to the "greenhouse effect."
- may cause severe burns.
- contributes to the erosion of our natural landscape.
- accelerates corrosion and rusting of many metals.
- may cause electrical failures and decreased effectiveness of automobile brakes.
- has been found in excised tumors of terminal cancer patients.

Contamination is reaching epidemic proportions!

Quantities of dihydrogen monoxide have been found in almost every stream, lake, and reservoir in America today. But the pollution is global, and the contaminant has even been found in Antarctic ice. DHMO has caused millions of dollars of property damage in the midwest, and recently California.

[1] Reproduced by permission.

Despite the danger, dihydrogen monoxide is often used

- as an industrial solvent and coolant.
- in nuclear power plants.
- in the production of styrofoam.
- as a fire retardant.
- in many forms of cruel animal research.
- in the distribution of pesticides. Even after washing, produce remains contaminated by this chemical.
- as an additive in certain "junk-foods" and other food products.

Companies dump waste DHMO into rivers and the ocean, and nothing can be done to stop them because this practice is *still legal*. The impact on wildlife is *extreme*, and we cannot afford to ignore it any longer!

The horror must be stopped!

The American government has **refused** to ban the production, distribution, or use of this damaging chemical due to its "importance to the economic health of this nation." In fact, the navy and other military organizations are conducting experiments with DHMO, and designing multi-billion dollar devices to control and utilize it during **warfare** situations. Hundreds of military research facilities receive tons of it through a highly sophisticated underground distribution network. Many store large quantities for later use.

It's not too late!

Act **NOW** to prevent further **contamination**. Find out more about this dangerous chemical. What you don't know can hurt you and others throughout the world.

In the interest of fairness to opposing points of view, giving equal time to each (just as is commonly done in TV news specials), we include the following response, as it appears on an Internet web site:

DHMO Now More Than Ever

by DAN CURTIS JOHNSON[2]

Perhaps you've heard of it: a colorless, odorless liquid; a powerful coolant and solvent; an easily-synthesized compound which is used by industry, the military, commercial operations, and even private individuals.

Yes, we are talking about hydrogen hydroxide, also known as dihydrogen monoxide, and we are here to tell you that what you've heard about DHMO is probably not the whole truth. There are forces out there, such as the Coalition to Ban Dihydrogen Monoxide, who would seek to legislate its use and availability, placing heavy limitations on it—and eventually, eradicating it entirely.

In the interest of fairness, we invite you to see their argument, and then we urge you to return here, to learn the truth. Their subversive agenda must not be allowed to prevail.

Hydrogen hydroxide is beneficial!

It has been shown that hydrogen hydroxide enhances the functionality, growth, and health of many forms of life—including humans!—and current research suggests that it has become an integral part of our planet's ecological balance.

Hydrogen hydroxide is environmentally safe!

Opponents of dihydrogen monoxide would have you believe that it is some kind of uber-toxin, that it wreaks caustic terror on anything

---

[2] Reproduced by permission.

it touches. This couldn't be farther from the truth; when handled properly, it enhances nature rather than destroys it, and even a worst-case scenario DHMO accident would be a trifle for the natural cycles of our world to handle.

Hydrogen hydroxide is benign!

The Coalition and others have popularized the label "dihydrogen monoxide" over the more chemically-accurate "hydrogen hydroxide" because they know how loaded the former name is. "Monoxide" has become synonymous with pollution, toxic gases, industrial waste—and while hydrogen hydroxide is sometimes a factor in these problems facing our world today, it is rarely the dangerous element.

Hydrogen hydroxide occurs in nature!

To hear its naysayers' descriptions, one would think hydrogen hydroxide was solely the product of industrial technology; that it came from years of research in clandestine labs. This is not the case! Hydrogen hydroxide has been a part of nature longer than we have; what gives us the right to eliminate it?

We need hydrogen hydroxide!

Don't let an uneducated and terror-stricken mob of fanatics railroad you into giving up your right to choose! Support the use and distribution of hydrogen hydroxide in your neighborhood, city, state, and country!

Some of us devious academics have given the DHMO spoof to our undergraduate students, asking them for written analysis and comment. We hoped they'd comment on the fraudulent arguments used (parodied) in it. The responses were dreadful. Even those who recognized it as a spoof, realizing that DHMO was water, didn't get the point of the assignment. A majority of them seemed not to realize it ***was*** *a spoof and responded with: "Why haven't the news media told us of this?" "Why isn't something being done about this?" "Is there a* *cover-up?" These people are easy targets for snake-oil salesmen and passionate zealots promoting causes. Experiences such as this inspired the following parody.*

The hazards of solar power[3]

by DONALD E. SIMANEK

Over the years I have heard and read much nonsense, to the point where nothing surprises me any more. Recently an item caught my eye, a pamphlet put out by the Citizens' Reactionary Alliance Concerned with Keeping the Environment Decent. It's a nice example of the style and "logic" common to propaganda pieces from many groups of alarmists and "aginers."

Many groups and individuals propose that our government spend tax money on research and development of systems to utilize solar power. They urge construction of vast solar energy collectors which would convert sunlight to electricity to supply our energy needs. They even advocate putting solar collectors on roofs of homes, factories, schools, and other buildings. Proponents of this technology claim that solar power would be safer and cleaner than coal, oil, or nuclear energy sources.

We view these proposals with alarm. Unscrupulous scientists and greedy promoters are hoodwinking a gullible public. We consider it rash and dangerous to commit our country to the use of solar power. This solar technology has never been utilized on

[3] Reprinted from *The Vector*, **3**, 2 (Dec. 1978) pp. 18–19.

such a large scale, and we have no assurance of its long-range safety. Not one single study has been done to assess the long-term safety of electricity derived from solar energy as compared to electricity from other sources.

The promoters of solar power cleverly lead you to believe that it is perfectly safe. Yet they conveniently neglect to mention that solar power is generated by *nuclear fusion* within the sun. This process operates on the *very same* basic laws of nuclear physics used in nuclear power plants and atomic bombs!

And what is the fuel from which this power is derived? It is hydrogen, a highly explosive gas (remember the Hindenberg?) Hydrogen is the active material in H-bombs, which are not only tremendously destructive, but produce deadly fallout. The glib advocates of solar power don't even mention these disturbing *facts* about the true sources of solar power. What else might they be trying to hide from us?

In addition to the *known* dangers cited above, what about the *unknown* dangers, which very well might be worse? When pressed, scientists admit that they do not fully understand the workings of the sun, or even of the atom. They even grudgingly admit that our knowledge of the basic laws of physics is not yet perfect or complete. Yet these same reckless scientists would have us use this solar technology even before we fully understand how it works.

Admittedly we are already subject to a natural "background" radiation from the sun. We can do little about that, except to stay out of direct sunlight as much as possible. The evidence is already clear that exposure to sunlight can cause skin cancer. But solar collectors would gather and concentrate that sunlight (which otherwise would have fallen harmlessly on waste land), convert it to electricity and pipe it into our homes to irradiate us from every light bulb! We would not be safe from this cancer-producing energy even in our own homes!

We all know that looking at the sun for even a few seconds can cause blindness. What long-term health hazards might result from reading by light derived from solar energy? Will we develop cataracts, or slowly go blind? *Not one* medical study has yet addressed itself to this question, and none are planned.

In their blind zeal to plug us in to solar power, scientists seem to totally ignore possible fire hazards of solar energy. Sunlight reaching us directly from the sun at naturally safe levels poses little fire threat. But all one has to do is *concentrate* sunlight, with a simple burning-glass, and it will readily ignite combustible materials. Who could feel safe with solar energy concentrators on their roof? Could we afford the fire insurance rates?

These scientists, and the big corporations which employ them, stand to profit greatly from construction of solar-power stations. No wonder they try to hide the dangers of the technology and suppress any open discussion of them.

Proponents of solar power present facts, figures and graphs to support their claim that solar power will become less expensive, as conventional fuel supplies dwindle and solar power technology improves. But even if this is true, what will stop the solar power equipment manufacturers and solar power companies from raising prices when they achieve a monopoly and other fuel sources disappear?

Of course every technology has risks. We might be willing to tolerate some small risk—if solar energy really represented a permanent solution to our energy problems. But that is not the case. At best, solar energy is only a temporary band-aid. Recent calculations indicate that the "Sun Will Go Out in Five Billion Years As Its Fuel Runs Out" (Source: newspaper headline). As that calculation was made a year ago, we now have only four billion, nine-hundred and ninety-nine million, nine-hundred and ninety-nine thousand, nine hundred and ninety nine years left during which we can use solar energy. Wouldn't it be better to put our human resources and scientific brains to work to find a safer and more permanent solution to our energy needs?

Afterword

It is now fashionable to be skeptical and suspicious of science and technology. We have a whole crowd of "aginers"—against pollution, nuclear power, water fluoridation, and genetic engineering. They are certain our brains are being scrambled by electromagnetic waves from our computer monitors and cellular phones, and that we are getting cancer from the electromagnetic fields of high power lines. They continually warn us of the dangers of practically everything we do. Yet some of these same folks who fear electromagnetic fields are now applying magnets to their bodies as healing therapy, and some even sleep comfortably in the warm electromagnetic field of their electric blankets.

I don't want to suggest that such concerns be totally dismissed. Still, there is a certain *style* to the aginer's polemics which, in its extreme forms, goes beyond the limits of science, logic and reason. One almost expects that activists will soon be warning us of the possible dangers of eating mashed potatoes.

The aginers have failed to consider some potential hazards. One is solar energy. Most of the aginers are *in favor* of that! One wonders whether they have even tried to find objections to it. Since they have neglected this, it seemed about time that someone took up the challenge of writing an essay on the dangers of solar energy. The essay above uses the aginers' methods of presenting arguments, their logic, and their persuasive style. Reasonable persons must disagree with the logic of this essay, but one has to admit that there is *some* truth in it.

When this essay was made available on the Internet, in a page linked under the heading "humor," it stimulated e-mail from folks who took it seriously! Some just declared the author was an idiot, others suggested that it must have been written by someone who obviously knew nothing about physics. Some of these who failed to "get the joke" claimed to be students of engineering, preparing for careers in environmental engineering. Many of these displayed their own ignorance of science and engineering, which one hopes will be eventually remedied by their education. One eleven-year-old girl patiently and politely explained the scientific mistakes in the essay, and her explanations indicated that she knew quite a few correct things about science. So the future of science and technology isn't hopeless.

The estimate of the future life of the Sun was just what was required to close the satire with the crowning absurdity—taking a newspaper headline as a fact of science, and treating an estimate of that sort as if it were exact to 1 year. If readers didn't recognize this piece was satire earlier, surely that closing paragraph should have clinched it.

Actually, current estimates suggest that in something like five billion years, give or take a few billion, the Sun will enter red giant phase and have expanded in size to about the present orbit of Mars, but the planets will have moved farther away also, and the situation in our solar system will be—well—vastly different than it is now, and probably inhospitable to our present way of life. The sun will then still have many more billions of years of remaining life, but extrapolations that far into the future are even more shaky.

But does it pollute the environment?

While driving in Pennsylvania, a family caught up to an Amish buggy. The owner of the buggy obviously had a sense of humor, because attached on its back was a hand printed sign: "Energy efficient vehicle. Runs on oats and grass. Caution: Do not step on exhaust."

Scraps from the landfill

A camping expedition got lost and spent several weeks hiking over snow-covered peaks, across flower-strewn alpine meadows and fording across sparkling, trout-filled streams. Finally they came upon an automobile junkyard flanked by a smelly mountain of trash tires. "At last," gasped the expedition leader, "civilization!"

Beautify America. Swallow your beer-cans.

10

MIND SCIENCES

Anybody that would go to a psychiatrist ought to have his head examined.

Samuel Goldwyn (1882–1974) U.S. movie producer

What is mind? No matter. What is matter? Never mind.

Thomas Hewitt Key (1799–1875)

Some historians of science think Freud was a secret cross-dresser. That may explain the often-heard references to "Freudian slips".

Two laboratory rats in the psychology lab were discussing their situation.

"I think I've got my researcher conditioned."

"How's that?"

"Well, every time I press this button, he gives me food."

I feel better already.

When uncertain,
When in doubt,
Run in circles,
Scream and shout.

A matter of definition

Much of what we call insanity may just be a passionate commitment to a delusion.

SCHIZOPHRENICS UNITE!

Lysdexia? What's Lesdyxia?[1]

Hidden meaning

A psychiatrist and a physicist pass each other on the campus. Each says to the other, "Good morning."

As they walk on, the physicist says to himself, "Cordial fellow."

The psychiatrist mutters to himself, "I wonder what he really meant by that?"

---

[1] Modified from R. S. Dietz and J. C. Holden *Creation/Evolution Satiricon.* The Bookmaker, 1987, p. 48.

Are *you* a problem thinker?

It started out innocently enough. I began to think at parties now and then to loosen up. Inevitably though, one thought led to another, and soon I was more than just a social thinker.

I began to think alone—"to relax," I told myself. But I knew it wasn't true. Thinking became more and more important to me, and finally I was thinking all the time.

I began to think on the job. I knew that thinking and employment don't mix, but I couldn't stop myself.

I began to avoid friends at lunchtime so I could read Thoreau and Kafka. I would return to the office dizzied and confused, asking, "What is it exactly we are doing here?"

Things weren't going so great at home either. One evening I had turned off the TV and asked my wife about the meaning of life. She spent that night at her mother's. I soon had a reputation as a heavy thinker. One day the boss called me in. He said, "Skippy, I like you, and it hurts me to say this, but your thinking has become a real problem. If you don't stop thinking on the job, you'll have to find another job." This gave me a lot to think about.

I came home early after my conversation with the boss. "Honey," I confessed, "I've been thinking..."

"I know you've been thinking," she said, "and I want a divorce!"

"But Honey, surely it's not that serious."

"It is serious," she said, lower lip aquiver. "You think as much as college professors, and college professors don't make any money, so if you keep on thinking we won't have any money!"

"That's a faulty syllogism," I said impatiently, and she began to cry. I'd had enough. "I'm going to the library," I snarled as I stomped out the door.

I headed for the library, in the mood for some Nietzsche, with NPR on the radio.

I roared into the parking lot and ran up to the big glass doors ... they didn't open. The library was closed.

To this day, I believe that a Higher Power was looking out for me that night.

As I sank to the ground clawing at the unfeeling glass, whimpering for Zarathustra, a poster caught my eye. "Friend, is heavy thinking ruining your life?" it asked. You probably recognize that line. It comes from the standard Thinkers Anonymous poster.

Which is why I am what I am today: a recovering thinker. I never miss a TA meeting. At each meeting we watch a non-educational video; last week it was "Porky's." Then we share experiences about how we avoided thinking since the last meeting.

I still have my job, and things are a lot better at home. Life just seemed . . . easier, somehow, as soon as I stopped thinking.

Conditioning

An MIT student spent an entire summer going to the Harvard football field every day wearing a black and white striped shirt, walking up and down the field for ten or fifteen minutes throwing birdseed all over the field, blowing a whistle and then walking off the field. At the end of the summer, it came time for the first Harvard home football game, the referee walked onto the field and blew the whistle, and the game had to be delayed for a half hour to wait for the birds to get off of the field. He wrote his thesis on this, and graduated.

Clinical notes

Just because I'm paranoid doesn't mean they aren't really out to get me.

Just because I'm schizophrenic doesn't mean I'm beside myself with concern about it.

Is it possible to be a closet claustrophobic?

What's the difference between a psychotic and a neurotic? A psychotic thinks two and two make five. A neurotic knows that two and two make four, but he just can't stand it.

Of course we must be open-minded, but not so open-minded that our brains fall out.

A neurotic is a self-taut person.

I'm schizophrenic; and so am I.

You're just jealous because the voices talk to me.

Only left-handed people are in their right minds.

Introspection.

The psychiatric hotline

Hello, Welcome to the Psychiatric Hotline.

If you are obsessive-compulsive, please press 1 repeatedly.

If you are co-dependent, please ask someone to press 2.

If you have multiple personalities, please press 3, 4, 5 and 6.

If you are paranoid-delusional, we know who you are and what you want. Just stay on the line so we can trace the call.

If you are schizophrenic, listen carefully and a little voice will tell you which number to press.

If you are manic-depressive, it doesn't matter which number you press. No one will answer.

The hidden brain damage scale

Of the many psychometric devices designed to measure the dimensions of human variation, the Hidden Brain Damage Scale stands alone as the only instrument capable of predicting a preference for pimento loaf. For this reason, and despite the sizable revenues that might accrue from the copyright, we offer the scale here for public consumption. It was written in a flurry of graduate school insight some years ago by Robin Vallacher (Illinois Institute of Technology), Christopher Gilbert (private practice, New Jersey) and Daniel Wegner (Trinity University, San Antonio, Texas). Although a true–false format is recommended, we have found that many test-takers opt for the response of getting tangled up in the drapery.

1. People tell me one thing one day and out the other.
2. I can't unclasp my hands.
3. I can wear my shirts as pants.
4. I feel as much like I did yesterday as I do today.
5. I always lick the fronts of postage stamps.
6. I often mistake my hands for food.
7. I'd rather eat soap than little stones.
8. I never liked room temperature.
9. I line my pockets with hot cheese.
10. My throat is closer than it seems.
11. I can smell my nose hairs.
12. I'm being followed by a pair of boxer shorts.
13. Most things are better eaten than forgotten.
14. Likes and dislikes are among my favorites.
15. Pudding without raisins is no pudding at all.
16. My patio is covered with a killer frost.
17. I've lost all sensation in my shirt.
18. I try to swallow at least three times a day.
19. My best friend is a social worker.

20. I've always known when to close my eyes.
21. My squirrels won't know where I am tonight.
22. No napkin is sanitary enough for me.
23. I walk this way because I have to.
24. Walls impede my progress.
25. I can't find all my marmots.
26. There's only one thing for me.
27. I can pet animals by the mouthful.
28. My toes are numbered.
29. Man's reach should exceed his overbite.
30. People tell me when I'm deaf.
31. My otter won't go near the water.
32. I can find my ears, but I have to look.
33. I'd rather go to work than sit outside.
34. I don't like any of my loved ones.

Psychiatric dictionary for the computer age

agoraphobia—will only visit home page
anorexia nervosa—refuses even microchips
arachnophobia—extreme fear of Web sites
binge-eating disorder—tries to choose everything from the menu
bulimia nervosa—booting and rebooting
cocaine addiction—constantly getting online
delusions of grandeur—wants to be an icon
depression—system is down
exhibitionism—likes to open Macintosh in front of others
multiple personality disorder—has too many interfaces
obsessive–compulsive personality—continually presses Control key
pathological lying—never uses Fax
rodentophobia—anxiety about using mouse
schizophrenia—constantly presses Escape key
voyeurism—attracted to Windows

Piaget's developmental stages

sensimotor stage: can repair cars.
preoperational stage: can be a nurse, helping to prepare surgeons for their work.

concrete operational stage: can be a construction worker.
formal operational stage: may attend black-tie dinners.

It's all in your mind

Woman: My husband thinks he's a chicken.
Psychiatrist: How long has this been going on?
Woman: Two years.
Psychiatrist: Why did you wait so long to seek help?
Woman: We needed the eggs.

There was a young lady in Ealing
Who had a peculiar feeling
That she was a fly
And wanted to try
To walk upside down on the ceiling.

Then her physicist husband came home and found her actually walking on the ceiling. "Don't you know that's against the law of gravity?" She fell to the floor and said angrily "You had to open your big mouth!"

He's not *that* crazy!

Two men are escaping from a mental hospital late one moonless night. When they reach the roof all they have to do is jump a fair distance across to the next building and they are home free. The first man, a psychotic, afraid of nothing, is willing to jump. The second man, however, is afraid of the dark, and is seriously considering returning to the hospital to deal with the issue.

The first man volunteers to jump across with the flashlight they brought with them, and then shine the light back across to the second man. "You can walk across on the beam of light and we will be away from here."

"You must think I'm crazy," replies the second man. "You'll turn the flashlight off when I'm halfway across, and I'll fall."

Technicolor dreams

"Doctor, I've been having fantastic, vivid dreams. They are filled with strange creatures: dragons, dinosaurs, wizards who cast spells, and even cute animals and bugs that talk. And they are always in brilliant color."

"Oh, yes. That's not uncommon. I've seen many cases before, and it's nothing to worry about. We call them 'Disney spells'."

Scraps from beneath the couch

The only difference between myself and a mad man is that I am not mad.

Salvador Dali (1904–1989) Spanish Surrealist painter

Ego: that quality that lets a person who is in a rut think he's in the groove.

Psychopath: a crazy road.

A cannibal started seeing a psychiatrist. He claimed he was fed up with people.

Multiple Personality.

Intuition is insight information.

You have reason to be concerned when you can hear the handwriting on the wall.

Minds are like TV sets. When they go blank, it's best to turn off the sound.

Sometimes my left hand doesn't know what my right hand is thinking.

I'm waiting for someone to write a book titled "Cure Yourself Of Addiction To Self-Help Books In Seven Easy Steps: The last self-help book you will ever need."

Psychiatrists say that one out of four people are mentally ill. Check three of your friends. If they are OK, you're the one.

Q: Why did Cleopatra refuse to see a psychoanalyst?
A: Because she was the Queen of Denial.

An Oedipal feline: a Freudy cat.

☞ According to popular books on psychology, the left cerebral hemisphere is supposed to carry out the logical and verbal thinking, while the right handles imaginative, creative and emotional thinking. This notion of a L/R split of cognitive functions is, however, too simplistic. Brain scans have shown that both halves of the brain participate in both kinds of process, and the two halves communicate quite a bit with each other during cognitive processing via the corpus callosum. Every mental activity is shared by both hemispheres, which differ mainly in processing styles. For example, people who have had right-brain damage due to strokes sometimes can understand literal language, but cannot grasp jokes, puns and allusions, and therefore should not buy this book. It does seem that the right brain excels in global, broad visual perception, while the left brain does better with fine details. Understanding of all of this is still incomplete and research studies sometimes produce puzzling and apparently contradictory results.

11

HISTORY OF SCIENCE AS IT WASN'T

Who says that humor isn't educational? Here's a collection of cartoons which deliberately misrepresent the history of science in various ways. The reader is invited to spot and correct the errors. It's a game of "What's wrong with this picture?" A few comments are included after each, to set the record straight.

Early man discovers that counting on the fingers can get you only so far in mathematics.

There's another problem here: the use of the symbol for zero on the signpost "10".

Place-value notation goes back at least to the Sumerians in Babylonia in the 18th century BCE, who wrote numbers in base 60 with cuneiform script. They had no zero *symbol*, however, merely leaving a space where a zero should be.

Indian mathematicians simplified the Babylonian number notation, replacing base 60 with base 10, and adapting it to decimal notation. Yet the Indians didn't realize the need for a symbol for zero until about 595 CE. The oldest European manuscript containing the symbols of this system dates from 976 CE, in Spain.

However, on the other side of the globe, the Mayans of Central America also invented a place-value system of notation, using base 20, which did include a symbol for zero.

If our finger-counting traveler decided to retrace his steps to zero, and then kept going, he'd be in the realm of negative numbers, another story entirely.

The first philosopher.

This is a classic logical paradox, warning us of the dangers of language, with its ability to express nonsense in a form which has a superficial appearance of profundity.

Our cartoon philosopher has just hit upon an insight so profound that he feels compelled to jot it down in permanent form for the enlightenment of future generations, not realizing that he thereby contradicts the very statement he's chiseling. Or is he making a statement about the multiple meanings and ambiguity of language? "Carved in stone" is a metaphor, not to be taken literally.

Suppose he rephrased the idea to avoid metaphor: "No statement is always true." This says that this particular statement is also not always true. Therefore one must conclude that there is at least one statement somewhere which *is* absolutely true, and that fact would contradict the premise. It would contradict it even if we were never able to find that absolutely true statement. This is an example of a class of self-referential statements which are capable of causing much havoc in logic.

"I have no idea what it's for. The Lord told me to construct an arc."

If the flood story in Genesis is to be taken seriously, a failure of communication like this would have had consequences which we wouldn't be here to contemplate right now.

Pythagoras explains his theorem.

The Pythagorean theorem, in modern algebraic notation tells us that the sum of the squares of the lengths of two legs of a right triangle is equal to the square of the length of its hypotenuse: $A^2 + B^2 = C^2$ where A and B are the lengths of the legs and C is the length of the hypotenuse. The legs are adjacent to the right (90°) angle, and the hypotenuse is the other side.

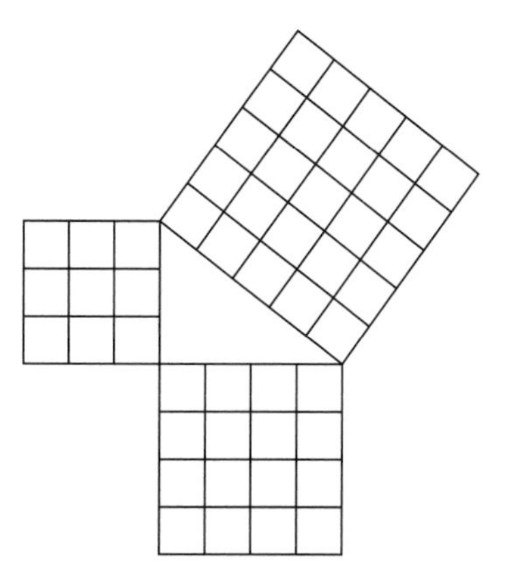

A 3:4:5 triangle, illustrating the Pythagorean theorem.

This theorem was independently discovered and known in various parts of the world. But Greek geometers were the first to see the need to prove its truth in a manner independent of mere measurement, and did so. The diagram above illustrates one of the special cases where the sides of the triangle are in integral ratio. In this case the integer lengths are 3, 4 and 5. But a special case, and a diagram, do not constitute proof of the general case.

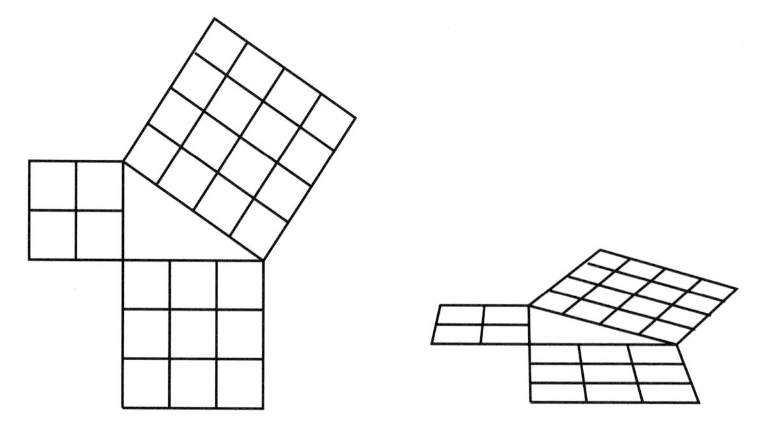

The mythical 2:3:4 right triangle, shown flat, and in perspective.

The drawing on the sand is subtly incorrect. It shows a 2:3:4 triangle, which isn't exactly a right triangle. Even when seen on the flat page it looks nearly correct!

Pythagoras' thought-balloon shows a rustic version of the Penrose impossible (illusory) triangle, not invented until the 20th century. We have no evidence that the ancient Greek mathematicians ever played around with illusory geometry. Yes, right triangles are mostly *hype*-otenuse. The hypotenuse is always longer than either of the legs of a right triangle.

The Penrose illusory triangle.

Why Galileo never dropped objects from the Leaning Tower of Pisa

Aristotelian academics in Galileo's day taught that heavy objects fall a given distance more quickly than lighter ones. A persistent legend tells that Galileo demolished this idea by dropping light and heavy balls from the Leaning Tower of Pisa. There is no solid evidence that Galileo ever did this. He did know of the experiment, for it had been done before: independently by the 7th century Byzantine scholar Joannes Philloponus (John the Grammarian), and also by the Flemish engineer Simon Stevin (1548–1620). In any case, the following spurious account is not the reason he didn't do the experiment.

> *On one visit to the Leaning Tower of Pisa he looked out from the balcony and realized that it would be a good place to try the experiment of the falling bodies. The Tower's inclination allowed a clear fall on one side of the Tower. But on that day he had neglected to bring along a clock to time the fall. The idea later slipped out of his mind and he never returned to perform the experiment. The moral of this story is that it's not sufficient to have the inclination if you don't have the time.*

Of course the flaw in this joke is the fact that there were no timing devices in Galileo's day capable of measuring such a short time interval. Other methods would have been used to determine which object reached the ground first, perhaps by letting them fall onto a wooden board and listening to determine which hit the board first.

On the basis of extensive and exhaustive research, and the dearth of reliable information, we lean toward the view that perhaps the evidence is still there for us to see. The Campanile Tower at Pisa leans. That's an undisputed fact. Galileo taught at Pisa, another undisputed fact. Did the Tower lean before Galileo came on the scene? Could it be that the Tower was perfectly straight until Galileo unwisely set up a crane anchored to the top of the Tower in order to lift a huge weight, and that effort caused the Tower to tilt? University officials would, of course, cover up this disaster. As evidence, we note that no university records of the experiment survive. This

"Really, Galileo, I think a much smaller weight would suffice to make your point."

could also explain why Galileo, embarrassed, never specifically mentioned the Tower of Pisa thereafter. It may even be that someone made up the story that the Tower was leaning previously, but no one noticed.

So much nonsense has been written about this alleged incident that we feel obligated to guide the reader to more reliable accounts. Lane Cooper published a pamphlet titled *Aristotle, Galileo, and the Tower of Pisa* (Ithaca, 1935; Kennikat Press, 1972) which should have set these many myths to rest. Still, textbooks continue to repeat the story, even including purely fabricated details, misrepresenting what Aristotle said about motion and misrepresenting what this sort of experiment is intended to demonstrate.

Galileo discovers the force of levity.

Aristotle thought that nature was mostly lawful, but sometimes behaved capriciously, allowing for miracles. Galileo is seen here being distracted from his experiment with falling balls as he notices an Aristotelian falling upward, in defiance of gravity. It must be "levity," the opposite of gravity! He doesn't even notice that the heavy ball has also fallen up, definitely a non-Aristotelian result. It's also an unphysical result.

One of the Alchemists' futile quests was to find the "universal solvent" which would dissolve **anything**. They never found it, so they never had to solve the problem of finding a container to store it in. It's a good thing, too, for such a material would be dangerous in the laboratory.

An unknown Alchemist discovers the two materials which, when mixed, produce the universal solvent.

René Descartes (1596–1650) was born in France, educated at the Jesuit College at La Flèche, studied law at Poitiers, then voluntarily served in the military (desiring travel and leisure time to think!). It must have worked. In 1619, during a winter in Germany with the army of the Duke of Bavaria, while sitting by a warm stove, he conceived a unified system of all knowledge, modeled on mathematics, based on physics and covering everything from medicine to morality. He claimed the whole work was rigorously rational. After traveling

Descartes finds it's not easy to get philosophy just right.

widely in Germany, Holland, Italy and France, he settled in Holland where he wrote his major works. In this picture we see him cogitating while the sea threatens to break through Holland's dikes.

Descartes' one (and only) lasting contribution to science was the Cartesian coordinate system as a basis for what is now known as analytic geometry. It allowed geometry problems to be translated into algebra problems.

Descartes' attempt to explain planetary motion was a muddled and complicated system of ether particles and vortices. For a while it attracted partisan advocates, but did not survive in competition with Newton's mechanics. His speculations on biology showed him to be out of his depth and unaware of the work of others in that field. His philosophy *was* more than the empty and meaningless slogan so often quoted "Cogito ergo sum" ("I think, therefore I am"). Yet very little of it has stood the test of time.

In 1649 Descartes made his biggest mistake. He accepted an invitation to tutor Queen Christina of Sweden in philosophy. The winters of Sweden, and the necessity of arising early for tutoring sessions, was too much for Descartes, who was used to reading in bed until 11 a.m. He died of pneumonia and is buried in Stockholm.

Newton gets some help in getting it right.

Correspondence between Newton and Nicolo Fatio de Dullier (a Swiss disciple, 20 years younger than Newton) suggests feelings characteristic of a romance.

Even Albert Einstein may have had trouble getting physics just right.

The often-quoted equation $E = mc^2$ is not something Albert pulled out of his mind to serve as a basis for his theory of relativity. It was an unexpected result which came out of a lengthy analysis after the theory of relativity was worked out rather fully.

The first application of the telescope was not to astronomy.

Historians aren't certain who first invented the telescope. Two persons, Janssen and Lippershey, both spectacle-makers in the Dutch town of Middleberg, independently claimed discovery of the telescope.

Johannes Lippershey (or Lippersheim) (15??–1619) of Middleburg applied for a telescope patent in 1608. Legend says this discovery was accidental. His young apprentice held one lens at arm's length and another lens near his eye, and saw a magnified image of the weather vane on a distant church steeple. If both lenses were converging, the image would have been upside-down. If the lens near the eye was diverging, the image would have been right-side up. We don't know what the apprentice was really looking for.

Turtle physics

In pre-scientific times the Hindus pictured the world as supported by elephants who had nothing better to do than to stand on the back of a huge turtle. We view such ideas with amusement today. They are interesting as history, literature and metaphor, but they are certainly *not* scientific models.

The eager student inquires of his master "What holds up the earth?" His wise teacher, steeped in the wisdom of the ancients, answers: "The earth is a flat disk supported on the strong backs of many elephants."

The student ponders this, then asks, "But what do these elephants stand upon?"

"They stand on the back of a very large turtle."

"But doesn't this turtle need something to rest upon?" the inquiring student asks.

"It sits on another turtle, in the very same manner as the first," the all-wise teacher replies.

"Well, then," the student says, "The same question arises as before: What does this turtle rest upon?"

"And the same logic must apply as before. The turtle sits on another turtle, which sits on another turtle, and from there on it's turtles all the way down."

The student is quite impressed by the consistency of this logic, but is still not satisfied. "Great master, the enormity of this system puzzles me. Please tell me, should not the same logic apply in the other direction? If each turtle has a turtle on its back, might there also be turtles all the way up? And what need is there for the elephants?"

The wise teacher smiled, and replied "You have much to learn, my boy. You have begun the infinite search for ultimate truth. You can only progress toward this truth by asking questions. But you will learn that each answer leads to another and more subtle question, which then leads to another. Great truths lie at each end of this chain of questions and answers, but they are ultimately no more profound than the chain itself."

When Archimedes was trying to "sell" folks on the idea of his law of the lever, he boasted "Give me a place to stand and I will move the earth!" But what would he stand upon?

Sir Isaac Newton supplied the metaphor when he said: "If I have seen farther than others it is because I stood on the shoulders of giants." Voila! The gedanken experiment is solved! Archimedes could stand on the shoulders of his philosopher-predecessors, Aristarchus, Euclid, Epicurus, Theophrastus, Aristotle, Plato, Socrates, Democritus, Zeno, Ptolemy... And it's philosophers all the way down. He would need more than something to stand on. He'd need a fulcrum as well. Perhaps an infinite stack of turtles would make a good fulcrum.

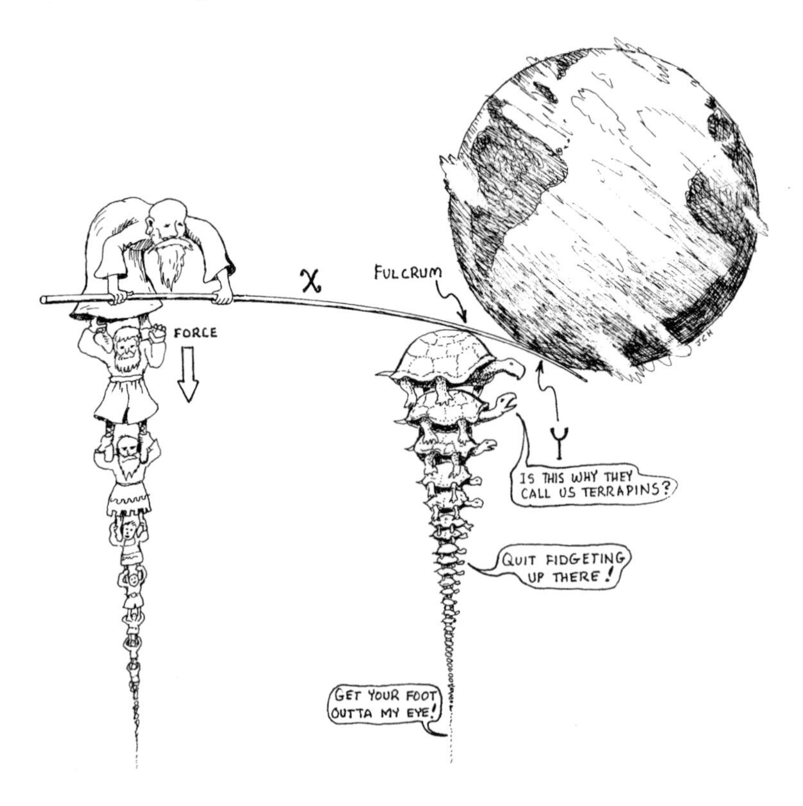

Problems

1. If the Earth's mass is 6.5×10^{21} lb and $Y = 100$ miles, how long would X be if Archimedes applied a force of 140 lb? [Note: The value of X may be expressed in parsecs.]
2. The fulcrum is composed of innumerable **terra**pins. ☐ True ☐ False.
3. Is this an example of animal abuse? Write a 3000 word essay and submit it to www.eureka!.org

---

Alfredo daVinci

Leonardo daVinci.

Leonardo daVinci (1452–1519) left us lavishly illustrated notebooks containing drawings of his studies for art works, as well as sketches of inventions, mechanisms, and notes on basic scientific studies. He was interested in anything mechanical, and his notebooks contain drawings of pulley systems, three of which are shown here (Figs 1–3).

These are not necessarily practical or useful systems, but at least they do work.

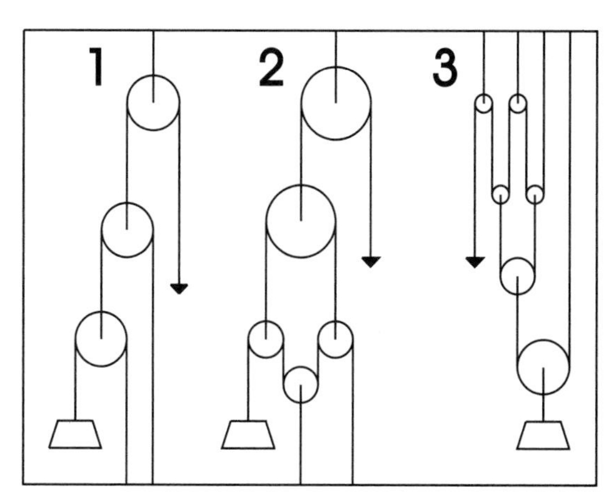

Figs 1–3. Pulley systems from Leonardo's notebooks.

Alfredo daVinci.

Unknown to historians, Leonardo had a brother, Alfredo, who wasn't nearly so clever as Leonardo. Alfredo continually tried, unsuccessfully, to compete with his older brother.

It is well known that Leonardo wrote his notebooks using "mirror" writing, which could be read by holding the page up to a mirror. Not to be outdone, Alfredo spent many laborious hours practicing, until he had perfected the ability to write both reversed right to left and upside down. He never figured out why other people had no difficulty reading it.

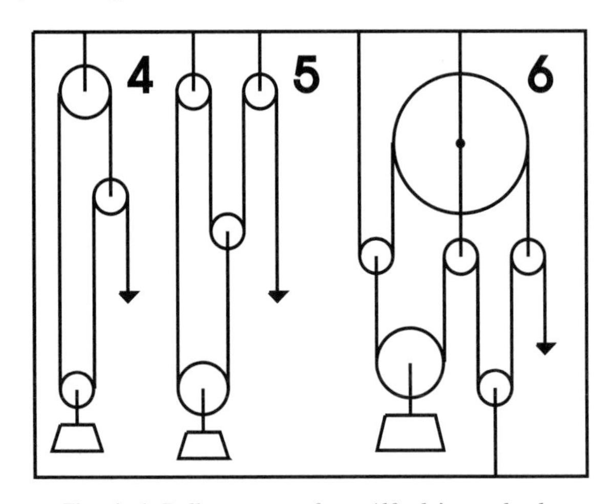

Figs 4–6. Pulley systems from Alfredo's notebooks.

Alfredo also kept notebooks, the contents of which have been universally ignored by scholars and historians. In them we find ample justification for his obscurity and the lack of recognition of his work. Figures 4–6 show three pulley systems from Alfredo's notebooks.

System 4 has come to be known as the "fool's tackle." Anyone who would try to set up one of these and make it work is foolish indeed. These systems have a frustrating tendency to collapse before they can be tested. A simple force analysis of the tension in the ropes reveals an internal inconsistency, that is, the tension forces simply don't add up. Therefore the system cannot be in equilibrium.

System 5 contains an inverted "fool's tackle," which works equally well. This illustrates that the basic fool's tackle, if inverted, or even if operated backward, still has the property of unworkability.

System 6 shows that it doesn't help to include two fool's tackles in the system, it just compounds the difficulties, and so this came to be called a "compound fool's tackle."

One as-yet-unproven fundamental theorem of classical physics asserts that any unworkable system of ropes and pulleys must have the basic fool's tackle somewhere within it, either normal, reversed, inverted or perverted.

Alfredo went on to devise even more innovative devices. You might say that he was continually re-inventing the square wheel. He is credited with the "blocked tackle," the "halted hawser" and the inverted pendulum. Alfredo must have realized how insignificant his inventions were, for he said "If I have not seen as far as others, it is because giants were standing on my shoulders."

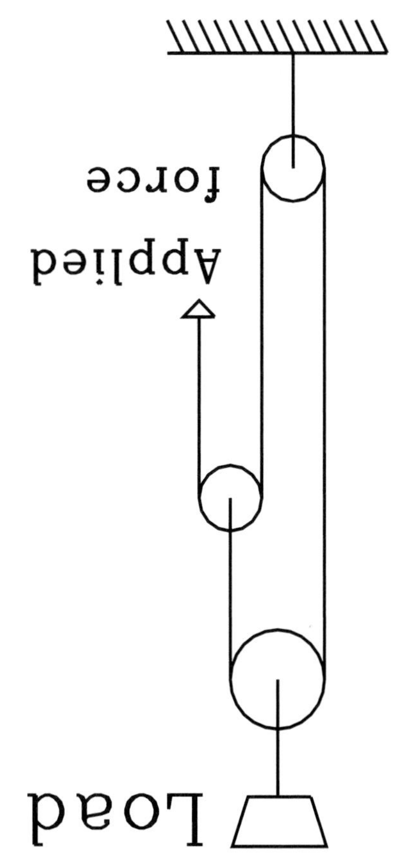

Alfredo's inverted pulley system, redrawn from his notebooks.

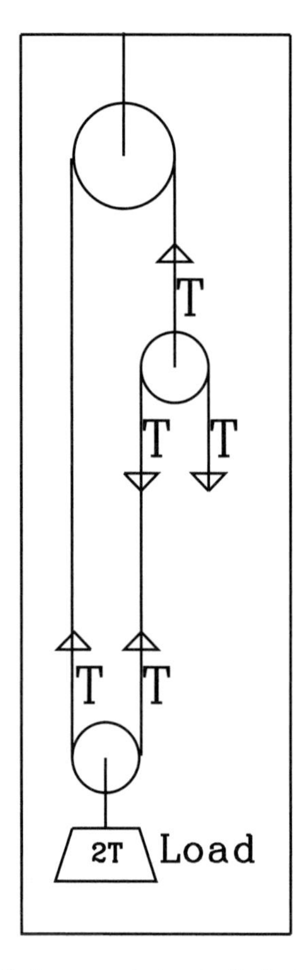

Fig. 8. Force analysis of a fool's tackle.

☞ Analysis of the normal fool's tackle: Consider the pulleys to be weightless and frictionless and the string to be perfectly flexible, inextensible and massless (to give us the best chance of making this thing work). Then the tension is uniform throughout the rope. That tension, T, is half the weight of the load. The bottom pulley is in equilibrium. But then the middle pulley has a net force of T downward, and cannot be in equilibrium. It will fall, and the system will be unusable. This shows that one may make drawings of systems which cannot physically exist as drawn. [Or this could be a snapshot of the system as it is collapsing!]

This is what happens when pulleys are arranged as a "fool's tackle". The invertability of sans-serif letters u, p, d, and n used to create an up/dn inversion was noted by Martin Gardner, and is used here with his permission.

12

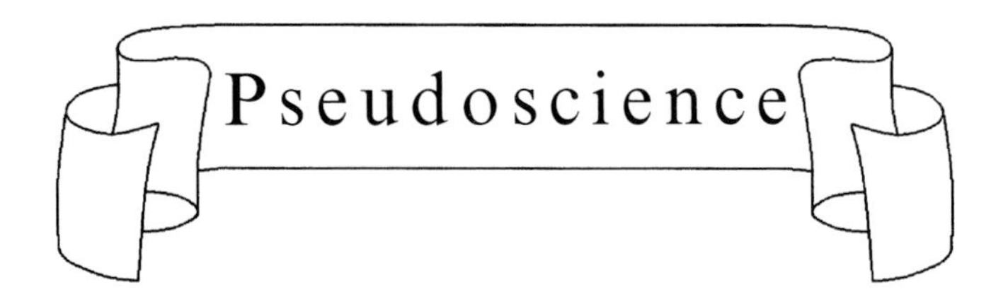

What is wanted is not the will to believe but the wish to find out, which is the exact opposite.

Bertrand Russell (1872–1970) English philosopher, mathematician, and writer

[Those] who have an excessive faith in their theories or in their ideas are not only poorly disposed to make discoveries, but they also make very poor observations.

Claude Bernard (1813–1878) French physiologist

Belief in the absence of compelling evidence is called faith. Belief after acquiring compelling evidence is called knowledge.

Carl Sagan (1934–1996) U.S. astronomer and writer

The mind abhors a vacuum. Without facts, they'll fill their heads with fantasies.

Jonathan Kellerman (1949–) U.S. psychologist and novelist
Time Bomb

The most costly of all follies is to believe passionately in the palpably not true. It is the chief occupation of mankind.

Henry Louis Mencken (1880–1956)
U.S. editor, critic and writer

Inspect every piece of pseudoscience and you will find a security blanket, a thumb to suck, a skirt to hold. What have we to offer in exchange? Uncertainty! Insecurity!

Isaac Asimov (1920?–1992) U.S. writer

Mankind's capacity for deception and self-deception knows no limits.

Paul Lee

The public will believe anything, so long as it is not founded on truth.
Edith Sitwell (1887–1964) English poet and critic

Prediction is difficult, especially when it involves the future.
Niels Bohr (1885–1962) Danish physicist.
Nobel Prize in Physics 1922

Like Granny always said, "If people could read thoughts, no one would get hit by a snowball."

Precognition

Some localities have laws against fortune-telling, palm reading, and other divination methods. Many of these fortune tellers have, over the years, been arrested and fined. Curious, isn't it, that they didn't see it coming.

Homo Sap

Phrenology, originated in 1798 by physician Franz Gall (1758–1828), held that the shape of one's skull reflected the development of portions of the brain associated with personality and character traits. Since this simplistic idea has been discredited, we update the classic phrenology head to represent what's really in the mind of the typical believer in pseudoscience, new-age mysticism and other such codswallop.[1]

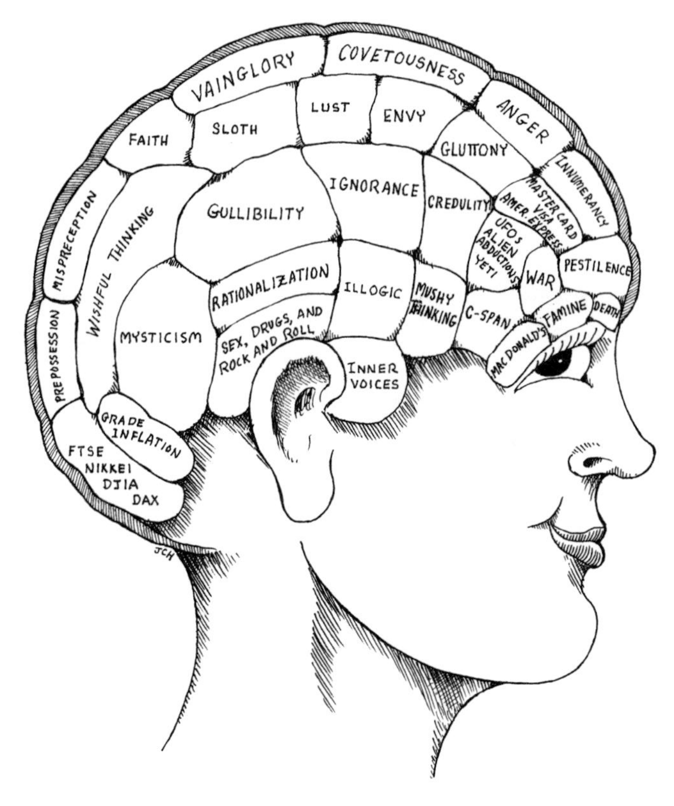

Homo Sap.

[1] Nonsense, talk or writing that is foolish or insincere.

Magnet therapy

Magnetic therapy promises new ways to improve life.

In stores and medical supply catalogs we see magnets in shoe inner soles, magnets in orthopedic braces and straps, even magnets in underwear. If you don't already have a magnetic personality, you can purchase one. These magnets are about the same strength

The healthful field of magnets.

as refrigerator magnets. If there is anything to this magnet therapy, your refrigerator, with all those magnets sticking on it, should be the healthiest place in the house. So go ahead; raid it frequently.

The snake-oil salesman

What if snake-oil salesman of a century ago were required to include disclaimers in their pitch? It might go like this:

From J. C. Holden and R. W. Hult **Now That You are Here . . . A Guide to the Methow Valley.** *The Bookmaker, 1976, p. 70.*

Let me introduce myself. I am Colonel Pompous A. Mountebank and I'm here today to offer you the once-in-a-lifetime opportunity to purchase my universal, miraculous, "Old Indian Snake Oil." This marvelous elixir has been compounded according to an authentic recipe handed down for generations by old Indians, who blended it from rattlesnake oil, bumbleberries, cactus roots, and secret herbs and spices found only in the hostile deserts of the great American southwest.

This marvelous product has been reputed to cure asthma, gout, rheumatism, prickly heat, shingles, catarrh, consumption, corns, catalepsy, croup, constipation, diarrhea and baldness. Each and every bottle carries an endorsement signed by Dr. Filbert Shyster, graduate *cum laude* of the College of Eclectic Medicine of Altoona, Pennsylvania. Of course, we paid him for the endorsement, as he was being assailed by several malpractice suits and needed the money.

This miracle medicine is manufactured the old-fashioned way in our own kitchen by my Indian assistant, Geronimo, and his wife Matilda. Well, actually, he's Indian only on his uncle's wife's side of the family, but he has thoroughly studied the lost Indian herbal lore passed down to him from his ancestors. A little later Matilda will show you an authentic Indian fertility dance, so stick around while I tell you more of this great opportunity to obtain what we all want: the hope and promise of lifetime good health.

Now we've tested this product on all sorts of maladies. When we get a cold, we take a tablespoonful each day and the cold goes away within a week. When we get a prickly rash, we rub it on and the rash is gone in a few days. Sore muscles relax after it's rubbed in with a good massage. If we take it every day as a preventative, we don't get a cold, or any other ailment worth complaining about, for many weeks running. So don't wait till disease or ailments strike you down and lay you low; take Colonel Mountebank's elixir every day.

Many satisfied customers tell us it cured their dropsy, senility, perspicacity, alacrity, verbosity and profundity. Now we don't know what all those ailments are, and shudder to imagine their symptoms, but we judge these folk to be honest individuals whose word we trust. That's why we paid them for their endorsements, printed on this inspiring brochure which my assistants are distributing to each and every one of you now.

We can't guarantee that this product will cure every ailment that afflicts you. You all know nothing's perfect. We don't make unrealistic claims that it will make you younger, or let you live forever. Results may vary. But the chance of an ailment or disease going away while taking this medicine is considerably greater than your chance of winning the Irish Sweepstakes, or any other lottery, and who among you hasn't bought a lottery ticket now and again? But to assuage all your doubts as to the wisdom of buying a bottle, or a case, or a year's supply of this wonderful medicine, we are offering you a money-back guarantee. If you can honestly say you were not benefitted, and do not even feel better after taking Mountebank's Snake Oil, we will refund your full purchase price upon return of the empty bottles and unused ones. Now as we are on the road much of the year, it may take some while for the Pony Express to find us, so allow at least a year for receipt of your refund.

Our humane method for making rattlesnake oil.

Some have asked where we find enough rattlesnakes for this elixir. I have to tell you that we gave up on rattlesnake oil as our business boomed and the rattlesnake population in our backyard declined catastrophically. We found, after intensive testing, that sunflower oil worked just as well, and was a lot safer to obtain, since sunflowers aren't usually vicious. But we did need to add just a touch of rattlesnake venom. We use a humane and ecologically sound method to obtain it. We distill some good Kentucky moonshine, and use it to attract the snakes. They drink it and become loopy and docile, so we can tease them to give up some venom in a bottle. This is later diluted by homeopathic methods. The snakes are released back into the wild no worse for the experience, and quite happy.

Our staff found that milkweed roots were as effective as cactus roots, and you don't have to go to the parched desert and brave sharp cactus spines to harvest them. Oh, yes, wild bumbleberries have become scarce as hen's teeth since Bob Evans' restaurants started serving a delicious dessert made from them. We found that blackberries combined with a homeopathic dilution of bumblebee venom works just as well. But I assure you that all of the ingredients are still blended according to the original old Indian recipe. We also guarantee that we use only natural and wholesome ingredients, never any man-made artificial chemicals. To ensure against spoilage, each bottle contains 100 proof genuine Kentucky moonshine as a preservative. As we said before, this product makes you feel better while it is curing whatever ails you.

Is Barney the evil one?

Everyone knows Barney, that cute purple dinosaur. But here's something that you may not know:

1. Start with the given:
   CUTE PURPLE DINOSAUR
2. Change all Us to Vs (which is proper Latin anyway)
   CVTE PVRPLE DINOSAVR
3. Extract all Roman Numerals:
   CV V L DI V
4. Convert these into Arabic values:
   100 5 5 50 500 1 5
5. Add these numbers up (or down if you wish):
   100
   5
   5
   50
   500
   1
   5

   666

There you have it, a numerically valid proof that Barney is the Antichrist!

☞ Numerology is a pastime and a kind of magical divination which claims to "deduce" significance from correspondences between names and numbers. If, by numerological calculation, two names give the same number, the persons with these names are supposed to be alike in some important way. The "number of the beast" mentioned in the *Book of Revelation* was 666. A popular numerological exercise was to use a person's name, nickname, initials, or whatever worked, to arrive at a result of 666, thereby "beasting" that person. With a little cleverness, using the full name, nickname or initials, and the right number–letter association, you can "beast" nearly anyone or anything.

The lighter side of the paranormal

A psychic researcher on a lecture tour was giving a public lecture in a small town in Scotland. Folks there are steeped in a tradition of haunted houses and ghosts, and so he had chosen "Spirits" as his topic. At one point, to elicit audience participation, he asked his audience "Has anyone here ever seen a ghost?" Quite a number of hands went up. "Has anyone here ever heard a ghost, even if it wasn't visible?" A similar number of folks indicated they had. "Has anyone here ever smelled a ghost?" A smaller number responded.

Then the speaker decided to inject a lighter note. "Has anyone here ever had sex with a ghost?" Only one hand went up, that of a rustic fellow in the back row. "Really! You have actually had sex with a spirit entity?"

"Oh, no!" the man replied. "I thought you said 'goat'."

Very latent spirit photography

Another psychic researcher was spending the night in an allegedly haunted house. He was awakened by a noise, and saw a ghostly fog seeping into the room through the keyhole. A spirit form materialized from the fog, and when it came into focus it was clearly seen to be that of a person, though its shape did shimmer and weave a bit. The researcher was astounded at his good fortune in meeting a ghost face to face for the first time.

Even more surprising, the spirit spoke. It was not threatening but actually friendly, glad to have someone to talk to. After some polite conversation, the researcher said, "Good Spirit, you are indeed a decent and friendly fellow. Would you mind if I take your picture? I have a camera here by the bed, and my friends will never believe I met you if I don't show them evidence."

To his surprise, the spirit readily agreed, even striking several poses in traditional spooky stance, while the researcher took flash pictures.

Eager to see the results, the researcher developed the film the next morning. But the pictures came out blank. You see, the spirit was willing but the flash was weak.

Heaven is bovine

A woman established contact with her late husband during a seance.

"Are you happy?" she asked.

"Oh, yes," came the answer. "It's a beautiful world here. The skies are blue; the pastures are green. And the females are gorgeous. They have gentle, loving eyes, and rounded forms."

The woman said, "I hope some day I can join you there in heaven."

"Heaven?" said the man. "Who said I was in heaven? I'm a bull in Montana."

Anti-Phlogiston?

You find the darndest things when you clean out a medicine cabinet—this jar of muddy-colored cream, for example. The label on the lid identifies it as an anti-phlogiston poultice. We knew that the phlogiston theory of 18th century chemistry was overthrown during the chemical revolution as a result of experiments done by chemists such as Lavoisier. Phlogiston, a substance supposedly exchanged during chemical reactions, was swept under the rug of science history and never taken seriously again. But if you still have inclinations to accept the phlogiston theory, just smear some of this cream on yourself and those feelings will go away.

Scraps from the spirit-cabinet

So you carry a rabbit's foot. Good luck. It didn't do the rabbit much good.

The ghosts who haunt skyscrapers are high spirits. A haunted elevator is a spiritual lift. Ghosts pick up their mail at the dead letter office. A haunted wigwam is a creepy tepee. The tales mediums relate from the dear departed are just seance fiction.

The aspiring witch never completed witches' school. She flunked the spell test.

A student from the Maharishi International University went to the dentist. He refused pain-killers, for he wanted to transcend dental medication.[3]

We are still waiting to hear about someone who has had a near-death experience in which they were on their way to Hell.

"I went to see a spiritualist."
"Any good?"
"No, just medium."

The afterlife is the one place you can go without having to pack a suitcase.

---

[2] The Maharishi University promotes the notion that meditation can develop one's ability to levitate, walk through walls and achieve omnipotence. A dentist related a true incident where a student from Maharishi university requested that all of her silver fillings be removed and replaced with gold. Her reason had nothing to do with concerns about the alleged toxicity of mercury in dental amalgam. She insisted that the silver interfered with her meditation.

13

THE EMPEROR'S NEW CLOTHES

A modern re-telling[1]

by A. SOP

The Mundavian empire was ruled, in its era of greatest power and prosperity, by a benevolent Emperor who, though possessing no great intellect, enjoyed the loyal and enthusiastic support of his people. He surrounded himself with advisors and courtiers of like mind who delighted in the simple pleasures of life: gourmet food, fine clothing, generous expense accounts, lavish parties and conspicuous decadence.

One fine day a merchant from the distant Orient arrived at the royal palace. He opened his cart and offered for sale a variety of exotic wares to the members of the Emperor's court.

After brisk sales of rugs, jewelry, and trinkets, the merchant announced that he was fortunate to have a limited supply of a new line of clothing: elegant creations fit for a King or an Emperor.

These garments, he said, had been produced by a small factory using a totally new principle of cloth manufacture. Unfortunately the firm had been absorbed in a hostile takeover. Clothing was dropped from the company operations and the secret of its manufacture was lost in locked corporate vaults. "This is a manufacturer's inventory close-out. When these are gone, there will be no more," the merchant assured his eager listeners.

---

[1] Astute and perceptive readers will find many references to science, pseudoscience and philosophy in this fable.

Virgins catching moonbeams.

When he sufficiently aroused his audience to the height of curiosity and expectation, an assistant brought in a large and heavy wardrobe trunk. With a flourish, the trunk was opened, but none present could yet see what was contained inside.

"These elegant garments can only be appreciated by those with the most sensitive discrimination and finest taste," the merchant said. "The designer has made a bold fashion statement by abandoning garish colors and florid ornamentation. Instead he's integrated disparate hues so subtle they can barely be perceived, using subliminal design details and fabrics so diaphanous they can barely be felt. Their sheer elegance was previously only imagined by less creative designers. The bold understatement of these ephemeral creations cannot be ignored."

He reached into the trunk to bring out a sample. "These fine pantaloons are made of the finest double-knit virgin polyunsaturated gossamer." He reached into the trunk again. "This robe is hand-woven by oriental craftsmen, using fiber spun from the purest moon-beams, harvested by oriental virgins on sparkling clear winter nights. These elegantly formal gowns are woven on air-looms capable of the most intricate designs."

"Air-looms?" the Emperor asked, with a puzzled look.

"Yes, looms driven by the wind—by air. The technology is revolutionary."

"Revolutionary?" The Minister of War looked alarmed.

"Metaphorically speaking, of course," replied the merchant, reassuringly.

The merchant carefully draped the robe over the emperor's shoulders. "See how light and sheer this fabric is. Wearing this,

A perfect fit.

you'll feel as if you have absolutely nothing on! This garment is so sheer that your body will be sensuously caressed by even the slightest breeze."

"Notice the subtle and delicate colors. The dyes were formulated by clever alchemists using principles lost forever by the ancients. These colors have enhanced radiation in the infraviolet and ultrared regions of the spectrum. They change hue and saturation to harmonize with your surroundings, appearing subtly different from every angle. Each garment in your ensemble will coordinate perfectly with any others. They go equally well with anything!"

"Of course you'll need footwear to complete your ensemble," the merchant said. "These shoes and slippers are made of leather from the wild Hypothesis, a beast so rare that none have ever been found. The leather is so supple that you'll imagine you are walking barefoot. And best of all, it's so soft and compliant that one size fits all." He slipped one on the emperor's foot.

"It's a perfect fit, don't you think?" The emperor wiggled his toes and nodded in agreement. "This footwear can't possibly pinch or squeak, never needs polish, and it is guaranteed never to leave ugly scuff marks on your palace's marble floors."

The court physicist was consulted, and asked if he knew what scientific principles made such garments possible. After thoughtful

The wild hypothesis.

deliberation he scribbled some equations on the back of a parchment envelope and ventured his conclusion that the clothes must be fabricated from quantum-mechanical forbidden states, linked into multidimensional strings of relativistic world-lines, through which light could tunnel without refraction. Indeed, he said, the uncertainty principle ensures that such garments should be undetectable by any scientific instruments, since any attempt to physically investigate them would result in their total destruction. This expert opinion duly impressed all who heard it, though none admitted to understanding it. The Emperor was pleased.

Some of those present could not, at first, see the magnificent garments. They were most perplexed by that fact, concerned that they might have a hitherto unnoticed deficiency of vision. Needless to say, they hesitated to admit this deficiency for fear of being considered insufficiently perceptive. The court soothsayer did venture to say that though he himself could not see the clothing, he did not in the least doubt its existence. He could only conjecture that some mysterious emanations from such magical clothing must be scrambling his sensory perceptions. This explanation was readily accepted and endorsed by those who were still having difficulty seeing the garments.

Much discussion ensued about the details of cut, styling and design. Some, who at first saw nothing, looked more carefully and found that they were gradually beginning to apprehend the clothing. They became convinced of this when they realized that the details they saw exactly matched those details which had been already described by others. It was only a matter of practice to improve one's observation skills. They were proud to have achieved this skill so quickly, feeling smugly superior to those who still saw nothing.

The court theologian was consulted, and asked whether wearing such clothing was morally acceptable. He pondered the matter carefully and for a considerable time, for this was a question which hadn't been covered in Divinity School. He also weighed the fact that his predecessor had been granted an early retirement to a cold, dark,

The court theologian is wary.

musty and rat-infested dungeon as a consequence of a hasty theological judgment. By now all those present claimed to perceive the clothing in all its magnificent detail. The initial difficulty some had experienced must have been due to a lack of faith. The only ones who might be morally corrupted by looking at persons so clothed would be those who hadn't enough faith to start with. Therefore, the theologian opined, so long as the styles were in good taste, there could be no religious objection to wearing the clothing. The merchant hastened to assure everyone that he offered no garment styles which might offend delicate sensibilities.

Sales were brisk. The merchant and his assistants were kept very busy describing the relative merits of the various garments, their color, style and cut. Fortunately the limited stock was exactly equal to the demand, and everyone was satisfied. Their huge chest, now empty of clothing, was filled to overflowing with gold and precious jewels received from the sales.

The court mathematician marveled at how all of those garments had come from so small a trunk. The merchant explained that the trunk had interior compartments specially constructed in an intricate geometric arrangement of cleverly intersecting null sets, thereby gaining storage capacity. This was a new concept for the mathematician, but he was confident that he could figure out the secret if he put his mind to it.

Word of the marvelous clothing soon spread among the populace, causing much speculation. Some (those of limited imagination) simply did not believe the stories, wanting to see the royal finery with their own eyes. The credibility of the Emperor and his government was seriously questioned in certain quarters. So the Emperor's advisors suggested that the new garments be shown to the public at the annual harvest festival. The entire court would attend, wearing the new clothing, for everyone to admire. That should satisfy the curious and settle all doubts.

The gala festival drew an unusually large crowd. The grand boulevard between the palace and the public amphitheater was lined on both sides with milling masses of people jostling for position to get a clear view of the royal procession. Everyone had high expectations of what they would see. Vendors wandered through the crowd selling parchments describing every detail of the clothing each royal dignitary would be wearing. Everyone eagerly anticipated the sumptuous display of finery they would soon be privileged to witness with their own eyes.

Finally their expectations were rewarded. The procession moved slowly down the boulevard, headed by the Emperor and his seven wives riding some of the finest horses in the empire. Other members of the court followed in ornately carved and jeweled open carriages. Though the horses and carriages were magnificently outfitted, the spectators agreed that these could not compare to the splendid and stylish robes of the emperor, the brilliant finery of the men, and the elaborate gowns of the women.

Public entertainments of this sort always attract vendors selling food, trinkets, funny hats and other novelties. One of these, a humble eyeglass-vendor, was doing a brisk business hawking special "Z-ray" glasses he claimed would enable anyone to see right through the new clothes of the Emperor's party. Of course most were skeptical of that claim, figuring that the glasses were just another fraudulent rip-off. Those who did buy them were also skeptical, but were willing to risk a few coins for the chance to see the royal personages in the buff. The vendor cautioned all buyers not to use the glasses until the Emperor stood up in his royal box at the amphitheater to officially pronounce the start of the day's festivities. Using them prematurely would, he said, void their warranty.

That eagerly anticipated moment arrived. The entire court stood before the multitude, proudly basking in the admiration of the crowd. The Emperor arose from his gilded chair to raise his royal scepter and let the festivities begin.

As he did so, many in the audience put on the special glasses. They were astounded at what they saw. "You can see right through their clothes!" a man cried. Someone else chimed in, "Look! They are naked underneath their clothes!"

Pandemonium ensued, but one of the Emperor's advisors quickly assessed the situation and figured out what was going on. An order was issued to apprehend the eyeglass vendor, who was captured as he attempted to flee the city.

He was charged with several counts of witchcraft and for selling enchanted glasses without an optometrists' license. For these crimes he was sentenced to death by being burned at the stake. He protested to the very end that the glasses were made of ordinary glass and had no occult power whatever.

The glasses were confiscated and destroyed. However, their magical power was so great that even without the glasses, people who had used them could *still* see through the royal clothing. They were even able to see through *anyone's* clothing! Some unfortunate individuals were so terrified by this evil enchantment

The moment of truth.

inflicted upon them that they sought help from priests, psychiatrists, and other charlatans who promised to purge the evil delusions from their minds. Occasionally the treatments worked. Others who came under the same spell, just smiled and said nothing.

And what of the wonderful clothing? For some while it became standard attire at court. Even those who couldn't see the clothing agreed that the finery enhanced royal balls, social intercourse, and even affairs of state. Visiting dignitaries were impressed with the finery, once it was described to them, and inquired where such clothing might be purchased. But though the Emperor sent his spies to the four corners of the earth, no one could find the cloth merchant, nor any other source of supply.

The following year the clothing was sent to the royal laundry for cleaning. The garments had not been cleaned before, for they had the remarkable property of not showing soiling. Still, the minister of health suggested that, at least once a year, laundering of any clothing was a good idea.

At the laundry all of the garments were accidentally lost. The Royal Launderer claimed that they must have dissolved in the wash water. In his defense he protested that he had searched for a "laundering instructions" label on the garments, but found none. He was, nonetheless, imprisoned for negligence. The members of the Emperor's court were therefore forced to return to more mundane clothing.

> ***Morals:*** *1. Delusions can be more persuasive than reality. 2. People who give up one delusion often replace it with another. 3. Those who see the reality beneath others' delusions are usually ignored or derided.*

Footnote

This was the official version of the incident, prepared by the Mundavian Minister of Historical and Political Correctness. Many years later, investigative reporters sought out more information about this event because of lingering suspicions that the eyeglass vendor and the Royal Launderer were innocent scapegoats. They found that amongst the spectators at the first public showing of the wonderful clothing was a young boy who had been unable to work his way to the front of the crowd to see the parade. A kindly man took notice of him. "Come up, boy, and sit on my shoulders," he said. "No one should miss this wonderful pageant of the Emperor's new clothes."

The boy was hoisted onto the man's shoulders and finally saw the royal personages who were proudly on parade. He could hardly believe his eyes. He pointed to the Emperor and exclaimed, "Why, the Emperor has no clothes!"

Spectators reacted in horror at this audacious outburst. But someone else in the crowd spoke up. "The boy is right. I can see it now. **None** of them have clothes!"

Others joined in. Soon nearly everyone could see the naked truth. "We were deceived by sorcery. Our minds were clouded, but this innocent boy has perceived matters as they really are."

The uproar reached the royal party. On all sides people were pointing and laughing, making suggestive and even lewd remarks about the royal personages they now perceived far more clearly than ever before. Chaos ensued as the members of the court realized their embarrassing condition. They scrambled to cover themselves with anything available: carriage curtains, horse blankets, and even parchment copies of government documents and Royal decrees.

Official accounts of this event attempted to divert attention from the gullibility and self-deception of the court. They pointed the finger of blame toward the innocent eyeglass vendor. This was probably the first historical example of a governmental "cover-up".

Morals: *1. Be suspicious of "official" versions. 2. Cover-ups are usually clumsily executed. 3. Be suspicious of the truth of any stories which claim to have a moral.*

14
MATHEMATICS

Insofar as mathematics applies to reality, it is not certain. Insofar as mathematics is certain, it does not apply to reality.

Albert Einstein (1879–1955) Swiss–U.S. physicist

Mathematicians are a species of Frenchman; if you say something to them they translate it into their own language and Presto! It is something entirely different.

Johann Wolfgang von Goethe (1749–1832)
German dramatist, novelist, poet

Two plus two equals five—for sufficiently large values of two.

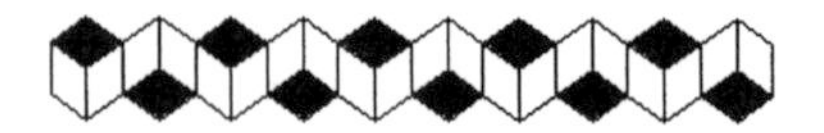

I nearly bought a book on infinity, but then I thought "No, I'll never have time to finish it."

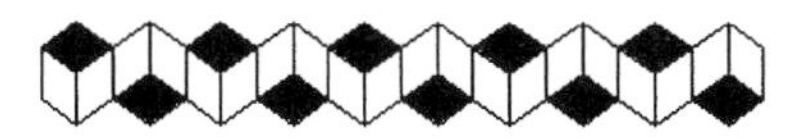

Life is complex: part real, part imaginary.

For every complex question there exists an answer that is simple, appealing and completely wrong.

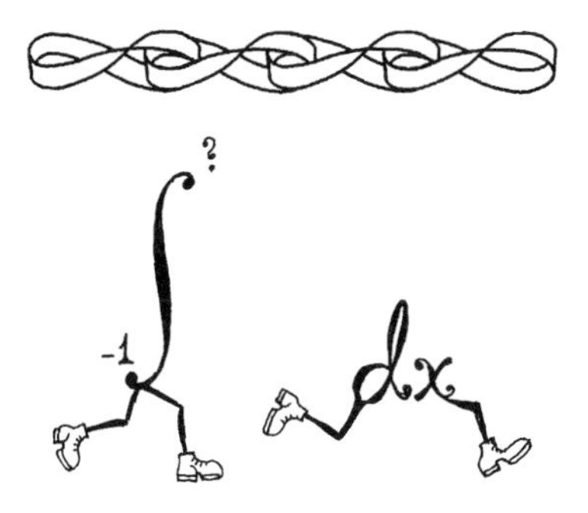

The integral sign and its differential remain together.

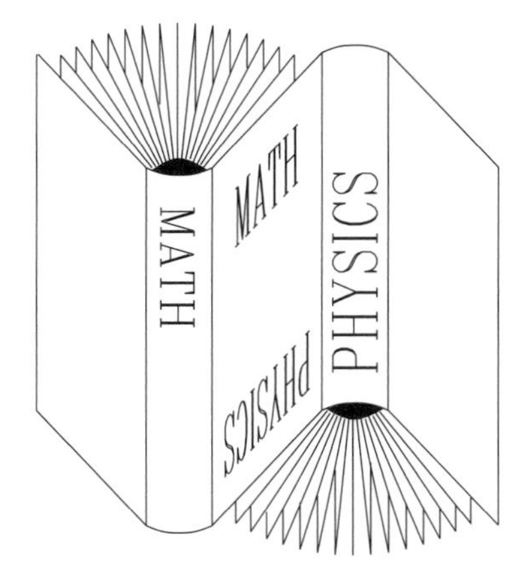

Mathematics and physics are inseparably symbiotic.

Anagrams

A decimal point.	I'm a dot in place.
One plus twelve.	Two plus eleven.
Integral calculus.	Calculating rules.

The proof is pudding

Readers who have taken calculus should be able to find the mistake in the following "proof."

Integrate

$$\int_1^2 \frac{1}{x}\,dx$$

Integrate by parts. Let $u = 1/x$ and $dv = dx$.

Then: $du = -\,dx/x^2$ and $v = x$.

Therefore

$$\int_1^2 \frac{1}{x}\,dx = 1 - \int_1^2 x\left[\frac{-dx}{x^2}\right]$$

So

$$\int_1^2 \frac{1}{x}\,dx = 1 + \int_1^2 \frac{1}{x}\,dx$$

Therefore

$$0 = 1 \quad \text{Q.E.D.}$$

How many students know what "Q.E.D" means? If you look it up in your Funk and Wagnalls, you'll find that it is an abbreviation for the Latin phrase "quod erat demonstrandum," which means "which was to be demonstrated." Students have other translations:

1. Quite easily done.
2. Quite erroneously done.
3. Question every deduction.
4. Quick, an English dictionary!
5. Quid erat Deus? (God, what was it?)

Then there's its cousin, Q.E.F., an abbreviation for "Quod erat faciendum," meaning "Which was to be done." Some math profs write a more modern abbreviation at the conclusion of an especially elegant proof, "W^4". That equals W·W·W·W which translates to: "Which is What We Wanted."

Oh, you are wondering where the mistake was? The formula for integration by parts is:

$$\int uv\,\mathrm{d}v = uv - \int v\,\mathrm{d}u.$$

This is a useful rule for evaluating indefinite integrals, but does not directly apply to definite integrals. When a change of variables is made in a definite integral the limits must also be re-expressed in terms of the new variable.

One plus one equals five, for sufficiently large values of one.

Base deception

A problem in number base conversions: Prove that Christmas = Halloween = Thanksgiving.[1]

Proof:

Christmas = DEC 25
Halloween = OCT 31
Thanksgiving = NOV 27

DEC 25 is 25 base 10 or $(2 \times 10) + (5 \times 1) = 25$
OCT 31 is 31 base 8 or $(3 \times 8) + (1 \times 1) = 25$
NOV 27 is 27 base 9 or $(2 \times 9) + (7 \times 1) = 25$

Certain impressionable minds might assume from this remarkable exercise in number-base conversions that there is some mystical significance relating the dates of these holidays.

[1] Solomon Golomb and John Friedlein, in Martin Gardner's *Mathematical Magic Show*, Alfred A. Knopf, 1977, p. 72 and 80.

The evolution of a mathematics problem

Arithmetic test, 1950: A logger sells a truckload of lumber for $100. His cost of production is four-fifths of the price. What is his profit?

1960: This has become subtly modified for students baffled by fractions: A logger sells a truckload of lumber for $100. His cost of production is four-fifths of the price, or $80. What is his profit?

1970 (The New Math): A logger exchanges a set L of lumber for a set M of money. The cardinality of set M is 100, and each element is worth $1.00. Make 100 dots representing the elements of the set M. The set C of the costs of production contains 20 fewer points than set M. Represent the set C as a subset of M, and answer the following question: What is the cardinality of the set P of profits?

1980: A logger sells a truckload of lumber for $100. Her cost of production is $80, and her profit is $20. Your assignment: find and circle the number 20.

1990: (Outcome-based education): "An unenlightened logger cuts down a beautiful stand of 100 trees to make a $20 profit. Write an essay explaining how you feel about this as a way to make money. Topic for discussion: How did the forest birds and squirrels feel?" (There are no wrong answers.)

1996: By laying off 40% of its loggers, a company improves its stock price from $80 to $100. How much capital gain per share does the CEO make by exercising his stock options at $80? Assume capital gains are no longer taxed because this encourages investment.

1997: A company outsources all of its loggers. It saves on benefits, and when demand for its product is down, the logging work force can easily be cut back. The average logger employed by the company earned $50,000, had 3 weeks vacation, received a nice retirement plan and medical insurance. The contracted logger charges $50 an hour. Was outsourcing a good move?

1998: A logging company exports its wood-finishing jobs to its Indonesian subsidiary and lays off the corresponding half of its US workers (the higher-paid half). It clear-cuts 95% of the forest, leaving the rest for the spotted owl, and lays off all its remaining US workers. The company claims that the spotted owl is responsible for the absence of fellable trees and lobbies Congress for exemption from the Endangered Species Act. Congress instead exempts the company from all federal regulation. Do you think the cost of lobbying paid off?

1999: A laid-off logger serving time in Federal Prison for blowing away his previous boss and a secretary is being retrained as a COBOL programmer in order to work on Y2K projects. What is the probability that his automatic cell doors will open on their own on 00:00:01, 01/01/00?

This account of the evolution of mathematics teaching has been widely circulated in several versions. Each year the rapid pace of educational change in math curricula inspires new additions. Our version is a melding of these. We've been unable to find the original, or identify the original author. However, we did run across this version which may have undergone separate evolution:

The Evolution of Mathematics Teaching

(How a math problem changed its look)

Up to the 1960s: A peasant sells a bag of potatoes for $10. His costs amount to 4/5 of his selling price. What is his profit?

In the early 1970s: A farmer sells a bag of potatoes for $10. His costs amount to 4/5 of his selling price, i.e., $8. What is his profit?

1970s (The New Math): A farmer exchanges a set P of potatoes with a set M of money. The cardinality of the set M is equal to $10 and each element of M is worth $1. Draw 10 big dots representing the elements of M. The set of production cost is comprised of 2 big dots less than the set M. Represent C as a subset of M and give the answer to the question: What is the cardinality of the set of profits?

1980s: A farmer sells a bag of potatoes for $10. His production costs are $8 and his profit is $2. Underline the word "potatoes" and discuss it with your classmates.

1990s: A farmer sells a bag of potatoes for $10.00. His production costs are 0.80 of his revenue. On your calculator graph revenue versus costs. Run the "POTATO" program on your computer to determine the profit. Discuss the result with the other students in your group. Write a brief essay that analyzes how this example relates to the real world of economics.

Correct $\neq$ useful

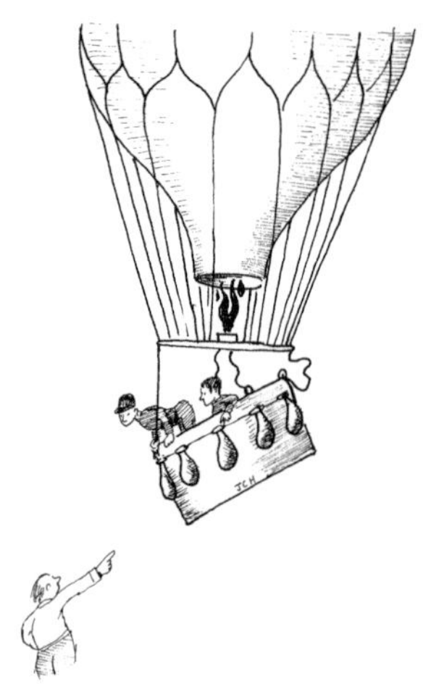

Two tourists took a balloon excursion somewhere in Europe. The balloon floated silently at a low altitude just above the treetops and rooftops of picturesque villages. They enjoyed the scenery, but hadn't brought a map. Then they spotted a fellow walking along the road one passenger leaned over the edge of the gondola and hailed him.

"Sir, can you tell us where we are?" he asked the man.

The fellow looked up, thought about the question, and answered "You are in a balloon."

"That didn't tell us anything," one passenger complained.

"He must be a mathematician," the other observed.

"Why? How can you know that?"

"Well, he thought a while before answering, then gave us an answer which was perfectly correct, precise, and of no use whatsoever."

Analog watches are still superior to digital watches in their failure mode. A stopped analog watch displays the correct time twice a day, a stopped digital watch is correct only once per day. And both of these are correct more often than a working clock which is running a bit fast or a bit slow. The running clock is correct less frequently as its rate is made more accurate. This is an example of the philosophical principle that a datum can be perfectly correct and still be absolutely useless.

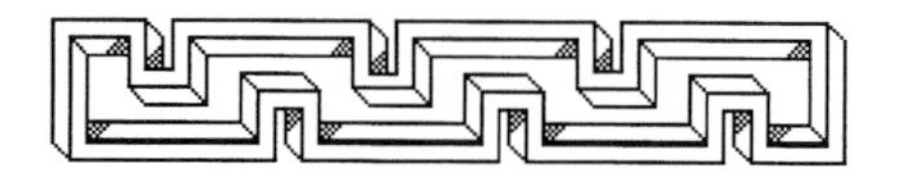

Native-American mathematics

Once upon a time in an Indian village, there lived three squaws. Two squaws had young sons. The first squaw placed her son, who weighed 60 pounds, on a deer hide near a pine grove; the second squaw put her son, who weighed 90 pounds, on a buffalo hide in the shade of a large oak tree; but the third squaw had no child. She rested on a hippopotamus hide beside a bubbling brook. Her weight was 150 pounds.

To this day, mathematicians give credit to these women and their children for demonstrating the Pythagorean theorem, because, you see: **The squaw on the hippopotamus equals the sons of the squaws on the other two hides.**

This joke circulates in many forms, always leaving the nagging question unanswered: How did an American Indian tribe obtain a hippopotamus hide? Perhaps they got it from a traveling circus whose hippopotamus died.

Be careful of assumptions

Two mathematicians having dinner were discussing the sorry state of mathematics literacy. One complains "The average person is a mathematical idiot. People can't do arithmetic correctly, can't balance a checkbook, calculate a tip or do percents." The other mathematician disagrees. "You exaggerate. People know a reasonable amount of math—all they need to know."

Later, when the complainer goes to the men's room, the other calls the waitress. He tells her he'd like to play a harmless joke on his friend.

"When my colleague returns in a few minutes, I want you to casually come by our table. I'll stop you and ask you a mathematics question. He'll suppose you won't know the answer. I'll give you the answer now, and he'll be amazed that you know it."

The waitress was mildly amused at this request. "So what's the answer?"

"One third x cubed."

The waitress asks "One thir dex cue"?

"Not quite". He repeats "One third x cubed".

"Oh, I've got it", she says with a smile. "One thir dex cubed"!

"Yes, that's good", he says. So she agrees to play her part, and goes off mumbling to herself, "One thir dex cubed . . .".

The other mathematician returns and his companion proposes a bet to prove his point that most people do know something about basic math. "I'm tired of your complaining. I'm going to stop the next person who passes our table and ask him or her an elementary calculus question, and I'll bet that person can solve it."

The skeptical mathematician figures its a safe bet, considering it unlikely that an ordinary person would be able to do calculus. The waitress comes back and is asked the question. "Can you tell me what is the integral of x squared?".

The waitress looks thoughtful for a bit, then replies "One third x cubed." The first mathematician is astounded. Then, as the waitress walks away, she turns and says over her shoulder "Plus a constant!"

Pictorial mathematics

Every calculus student learns that one must include the constant when evaluating an indefinite integral. A joke integral is of the form:

$$\int \frac{\mathrm{d}(cabin)}{cabin} = ?$$

The usual answer "log (*cabin*)" isn't quite right. The answer is "log (*cabin*) + C." One must include the constant "C", which gives you "A log cabin by the sea," or, perhaps, a "houseboat". Another old joke is the integral $\int e^x = f(u^n)$. But this is defective mathematics, for the integrand is missing the essential differential "dx".

Poetry in mathematics

Who says there's no poetry in mathematics? Here's a limerick in the form of an equation.

$$\left[\int_1^{\sqrt{3}} z^2 \mathrm{d}z\right] \cos\left(\frac{3\pi}{9}\right) = \ln \sqrt[3]{e}$$

Which, of course, translates to:

Integral z-squared dz
From 1 to the square root of 3
Times the cosine
Of three pi over 9
Equals log of the cube root of e.

If that was a bit over your head, here's a non-calculus math limerick:

$$\frac{12 + 144 + 20 + 3\sqrt{4}}{7} + (5 \times 11) = 9^2 + 0$$

A dozen, a gross, and a score,
Plus three times the square root of four,
Divided by seven
Plus five times eleven
Is nine squared and not a bit more.

What is "pi"?

Mathematician: Pi is the number expressing the relationship between the circumference of a circle and its diameter.

Physicist: Pi is 3.141 592 7 plus or minus 0.000 000 05.

Engineer: Pi is about 3.

Several students were asked the following problem: Prove that all odd integers are prime.

The mathematics student says "Hmmm . . . Well, 1 is prime, 3 is prime, 5 is prime, and by induction, we conclude that all the odd integers are prime."

The physics student then says, "I'm not sure of the validity of your proof. I'll try to prove it by experiment." He continues, "Well, 1 is prime, 3 is prime, 5 is prime, 7 is prime, 9 is . . . uh, 9 is an experimental error, 11 is prime, 13 is prime . . . Well, it seems that you are right."

The engineering student responds, "Actually, I'm not sure of your answer either. Let's see. 1 is prime, 3 is prime, 5 is prime, 7 is prime, 9 is . . . , 9 is . . . , well if you approximate, 9 is prime, 11 is prime, 13 is prime . . . Yes, it does seem correct."

Not to be outdone, the computer science student says "Well, you two have the right idea, but you'd end up taking too long doing it. I've just whipped up a program to *really* prove it . . ." He goes to his computer and runs his program. Reading the output on the screen he says, "1 is prime, 1 is prime, 1 is prime, 1 is prime . . . ."

☞ A serious flaw in this joke was noted by our in-house mathematician/proofreader. In mathematics the term "prime number" is defined to be "any integer which has no integral factor other than 1 and itself. Usually 1 is excluded from the list, and sometimes 2 is excluded also, since it is the only even prime and therefore it would be the oddest prime". We wouldn't want math students to be misled by this very widely circulated joke.

A biologist, a statistician, a mathematician, a computer scientist, and a philosopher are on a photo-safari in Africa. They drive out on the savannah in their jeep, stop and scout the horizon with their binoculars.

The biologist: "Look! There's a herd of zebras! And there, in the middle: A white zebra! It's fantastic! There *are* white zebras! We'll be famous!"

The statistician: "It's not significant. We only know there's one white zebra."

The computer scientist: "Oh, no! A special case!"

The mathematician: "Actually, we only know there exists a zebra, which is white on one side."

The philosopher: "We can't really say for certain that zebra has another side."

An assemblage of the most gifted minds in the world were posed the following question: "What is 2×2?"

The engineer whips out his slide rule (yes, this is an old joke) and shuffles it back and forth, and finally announces "3.99".

The physicist consults his technical references, sets up the problem on his hand calculator, and announces "It lies between 3.999999 and 4.000001".

The mathematician furiously writes equations on a scratch pad, oblivious to the rest of the world, then announces: "I don't know what the answer is, but I can prove to you that an answer exists!"

The philosopher cogitates for a while, then asks "But what do you *mean* by 2×2?"

The logician says "Please define "2" and "$\times$" more precisely."

The accountant closes all the doors and windows, looks around carefully, then asks "What do you *want* the answer to be?"

Taming the wild equations

Wild equations can sometimes be tamed. Consider this magnificent example:

$$\ln\left[\lim_{z \to \infty}(1 + z^{-1})^z\right] + (\sin^2 x + \cos^2 x) = \sum_{n=0}^{\infty} \frac{\cosh y(1 - \tanh^2 y)^{1/2}}{2^n}$$

Application of elementary algebra and trigonometry relations shows that this wild equation reduces to the lowly tautology $1 = 1$. This, of course, is normally left as an exercise for the student.

Here's how it's done. Apply the elementary trigonometric identity

$$\sin^2 x + \cos^2 x = 1,$$

and the well-known relation for hyperbolic trig functions

$$\cosh y(1 - \tanh^2 y)^{1/2} = 1.$$

Evaluate the limit

$$\lim_{z \to \infty} (1 + z^{-1})^z = e.$$

Don't forget that

$$\ln e = 1.$$

Now sum the series

$$\sum_{n=0}^{\infty} \frac{1}{2^n} = 2.$$

Finally, divide both sides by 2. The result is the tautology $1 = 1$.

Approximately ten excuses for not doing math homework

- I accidentally divided by zero and my paper burst into flames.
- I could only get arbitrarily close to my textbook. I couldn't actually reach it.
- I have the proof, but there isn't room to write it in this margin.
- I was watching the World Series and got tied up trying to prove that it converged.
- I have a solar powered calculator and it was cloudy.
- I locked the paper in my trunk but a four-dimensional dog got in and ate it.
- I couldn't figure out whether I am the square of negative one or I am the square root of negative one.
- I took time out to snack on a doughnut and a cup of coffee, and then I spent the rest of the night trying to figure which one to dunk.
- I could have sworn I put the homework inside a Klein bottle for safekeeping, but this morning I couldn't find it.

Residues

A mathematician is a device for turning coffee into theorems.
Paul Erdös (1913–1996) Hungarian mathematician

Mathematicians' bakery: House of Pi's.

Zenophobia: The irrational fear of convergent sequences.

Algebraic symbols are used when you don't know what you are talking about.

Theorem: A cat has nine tails.
Proof: No cat has eight tails. A cat has one tail more than no cat. Therefore, a cat has nine tails.

Mathematics is made of 50 percent formulas, 50 percent proofs and 50 percent imagination.

Ralph: Dad, will you do my math for me tonight?
Dad: No, son, it wouldn't be right.
Ralph: Well, you could try.

15

STATISTICS IS A CHANCY BUSINESS

There are three kinds of lies: Lies, damn lies and statistics.
Benjamin Disraeli (1766–1848) English author and politician

Statistician: One who collects data and draws confusions.
Hymon Berston

Statistics: A group of numbers looking for an argument.
Anon

Lake Wobegon, . . . where . . . all the children are above average.
Garrison Keillor (1942–) U.S. humorist and radio raconteur

Facts are stubborn, but statistics are more pliable.
Mark Twain (1835–1910) U.S. author and humorist

One of the most untruthful things possible, you know, is a collection of facts, because they can be made to appear so many different ways.
Dr. Karl Menninger (1893–1990) U.S. psychiatrist

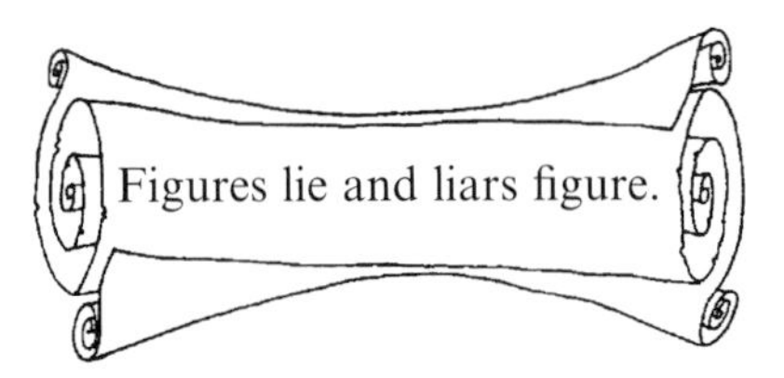

I'm never quite sure about uncertainties.

Statistics means never having to say you are certain.

Being average is a mean accomplishment.

A statistician says that figures don't lie, but must admit they don't always stand up.

All probabilities are 50%. Either something happens, or it doesn't.

According to recent surveys, 51% of the people are in the majority.

50% of the population have a below-average understanding of statistics.

It always pays to check your work

A student was completely hung over the day of his final statistics exam. It was a true/false test and he knew he'd be unable to even comprehend the questions, so he decided to determine the answers by flipping a coin. The professor watched the student the entire time as he was flipping the coin . . . writing the answer . . . flipping the coin . . . writing the answer. At the end of the two hours, everyone else had left the final except for that one student.

The professor walked up to the student and said "Listen, it's obvious to me that you did not study for this statistics test. You didn't even open the exam booklet. If you are just flipping a coin for your answers, what is taking you so long?"

The student replied impatiently (still flipping the coin): "Shhh! I'm checking my answers!"

Dicey dice

Before taking a chance with dice, it's a good idea to examine them carefully. The dice in the illustration on the previous page are marked improperly. Other ideas for mis-marked dice:

(a) Some people are born winners.
(b) Some people are born losers.
(c) Some people are born *tabula rasa*. These are also known as "Christian dice" for those whose religious principles prohibit gambling.
(d) Heisenberg die. Only one is needed to play.
(e) Maybe these are fair, but you'd better look at the other three faces.

He uses statistics as a drunk uses a lamppost; for support rather than illumination.

Andrew Lang (1844–1912) Scottish author and scholar

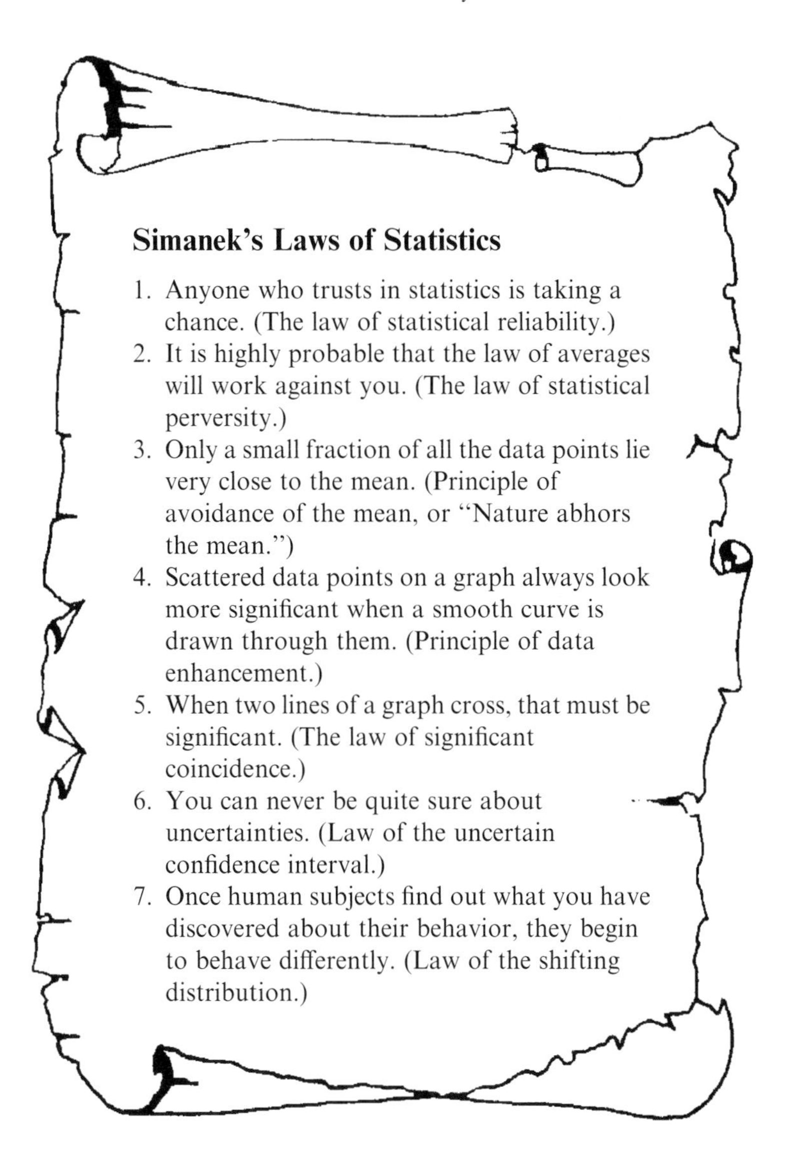

Simanek's Laws of Statistics

1. Anyone who trusts in statistics is taking a chance. (The law of statistical reliability.)
2. It is highly probable that the law of averages will work against you. (The law of statistical perversity.)
3. Only a small fraction of all the data points lie very close to the mean. (Principle of avoidance of the mean, or "Nature abhors the mean.")
4. Scattered data points on a graph always look more significant when a smooth curve is drawn through them. (Principle of data enhancement.)
5. When two lines of a graph cross, that must be significant. (The law of significant coincidence.)
6. You can never be quite sure about uncertainties. (Law of the uncertain confidence interval.)
7. Once human subjects find out what you have discovered about their behavior, they begin to behave differently. (Law of the shifting distribution.)

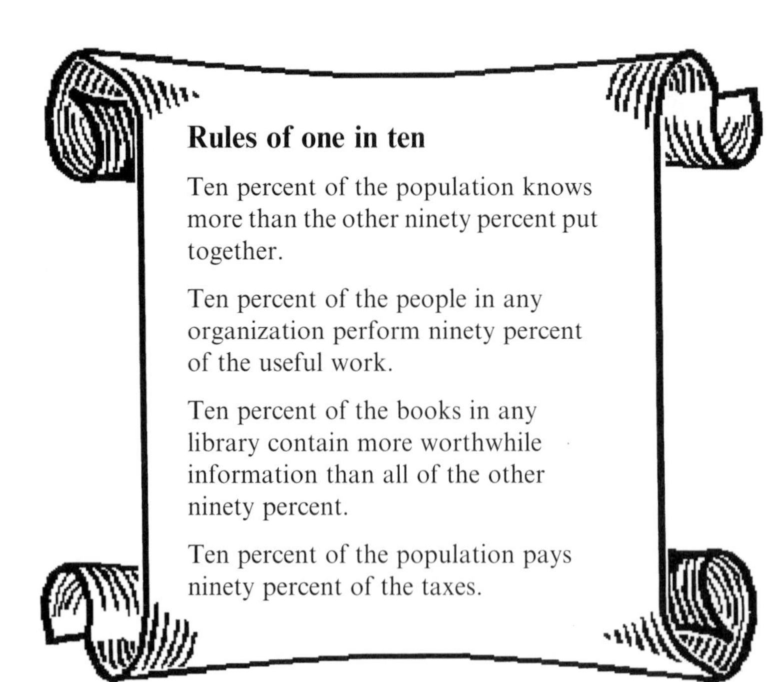

The New England Journal of Medicine reports that 9 out of 10 doctors agree that 1 out of 10 doctors is an idiot.
James (Jay) Leno (1950–) U.S. comedian and TV host

Ten percent of the population gets into ninety percent of the trouble.

Only 10 percent of anything can be in the top 10 percent.

If there is 50–50 chance that something can go wrong, then nine times out of ten it will.

Rules for graphing experimental data

Graphs can be powerful communication tools. Unfortunately the message you wish to communicate may sometimes be only weakly (or not at all) supported by the data. There are, however, creative ways to make the graphs say what you want them to say, rather than what the data indicates.

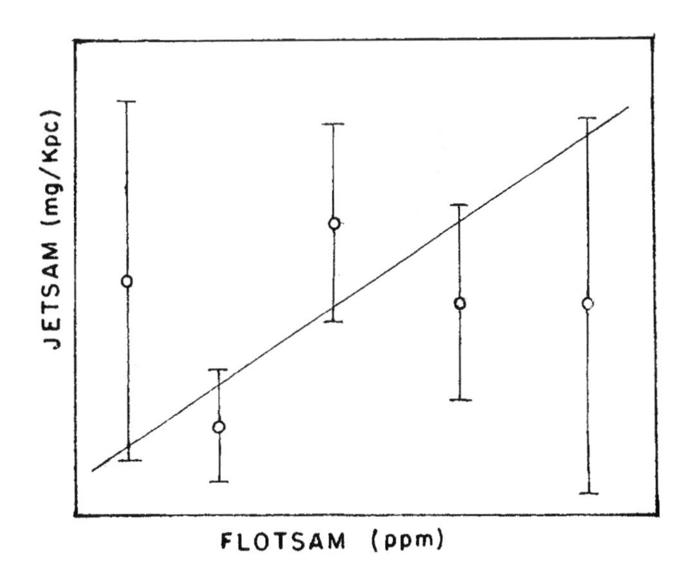

Draw the error bars large enough to include the function you want to demonstrate.

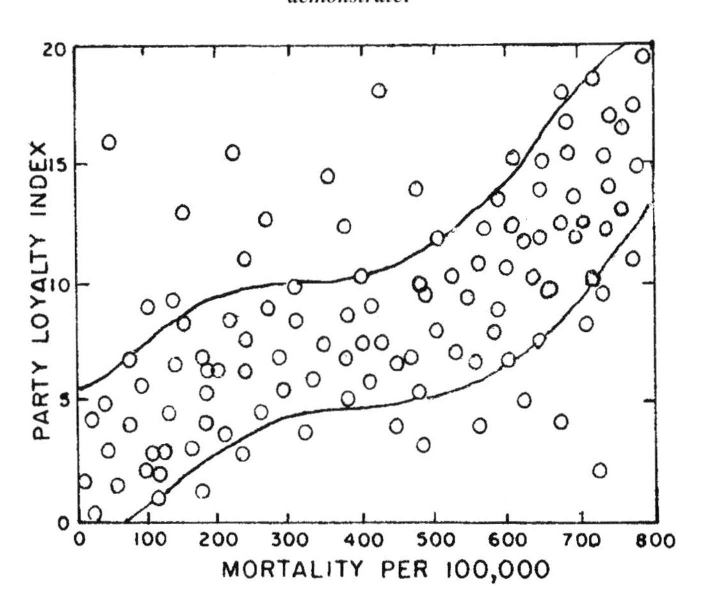

Make the width of the regression line proportional to the precision of the data.

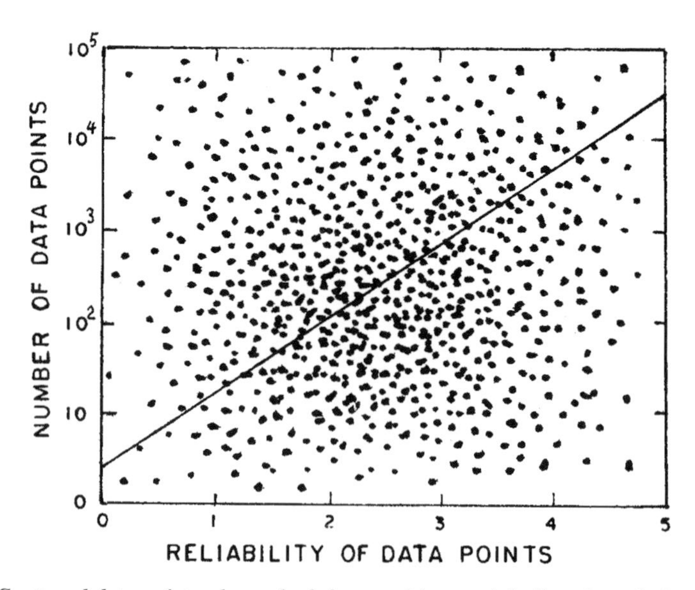

Scattered data points always look better with a straight line through them.

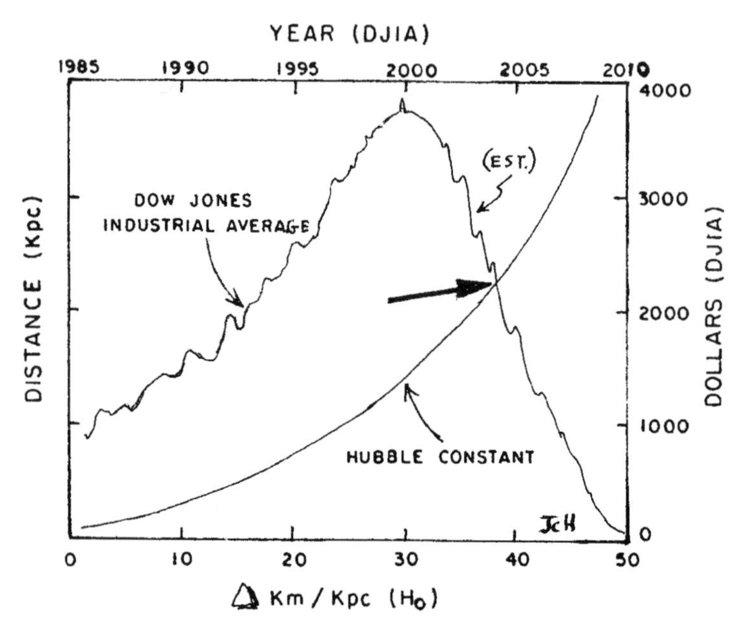

If two lines on a graph cross, it must be important.

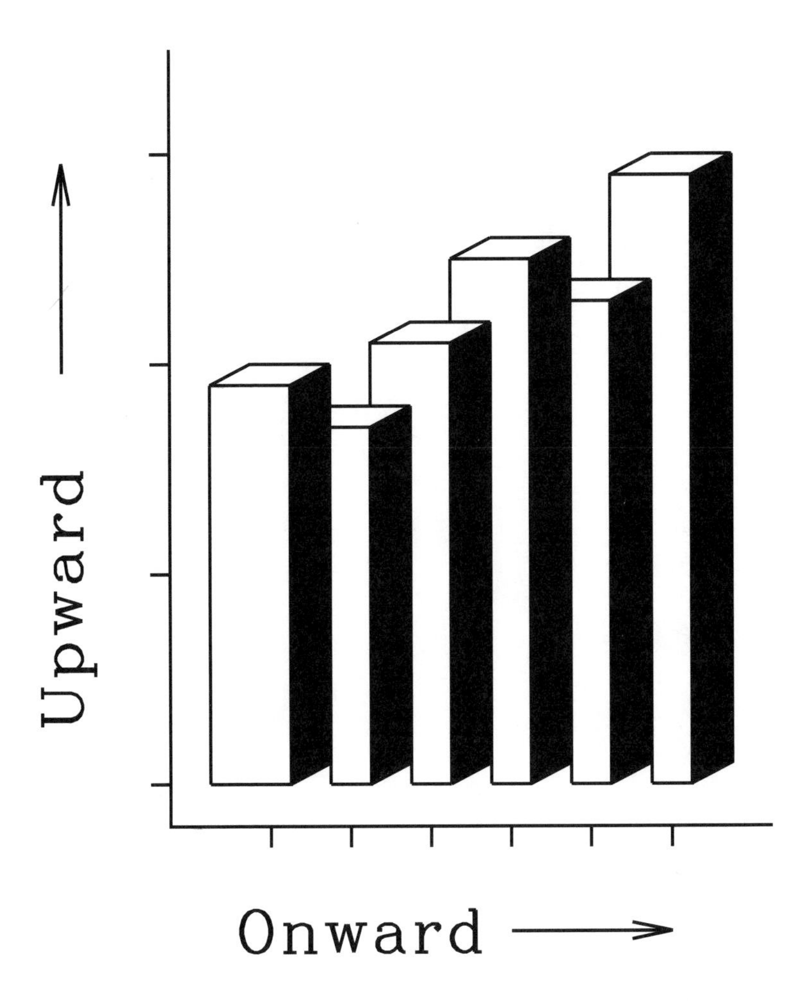

Bar graphs are popular in business presentations, for they dramatize the conclusion you wish to get across, while distracting your audience from irrelevant issues such as the validity of the underlying facts and alternate conclusions. Of course it's unprofessional to use anything but full color.

The example shown below demonstrates how the impact of a graph may be enhanced by artistic embellishment and "humanizing" touches. This style of presentation can be a work of art.

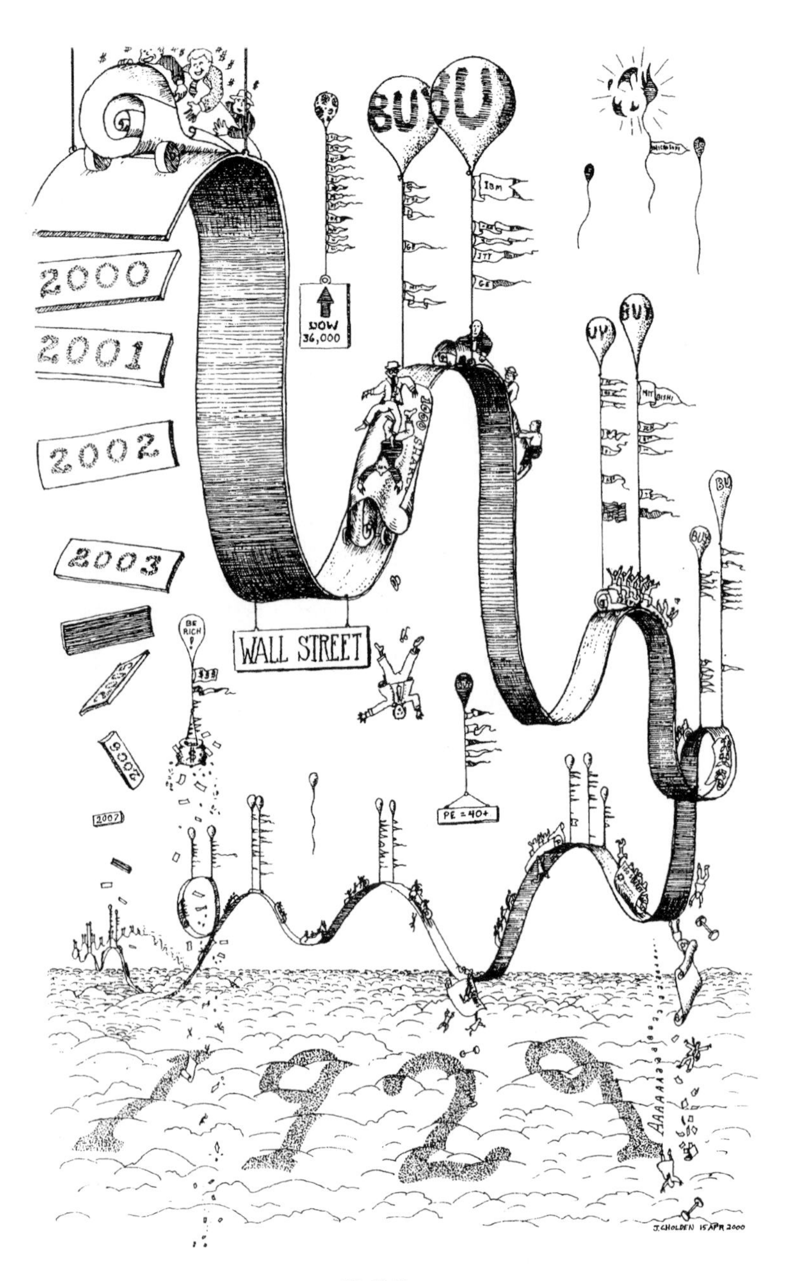
2000
2001
2002
2003
WALL STREET
DOW
36,000
BE RICH !
PE = 40+
IBM
J. CHOLDEN 15 APR 2000

Wall Street

Important decisions should be based on sound statistical principles.
From A. E. Benthic* The Id of the Squid. *Compass Publications, 1970, p. 68.

Statistics helps us understand the real world

Statistics show that:

Nearly two out of three families live next door.

The chief cause of divorce is marriage.

Marriage is 83% effective in preventing suicide. On the other hand, statistics show that suicides are 100% effective in preventing subsequent marriage.

67.3% of statistics are made up.

Infertility is hereditary. If your parents had no children, chances are you won't either.

Insanity is hereditary. You get it from your children.

Birthdays are good for you. People who have the most birthdays live the longest.

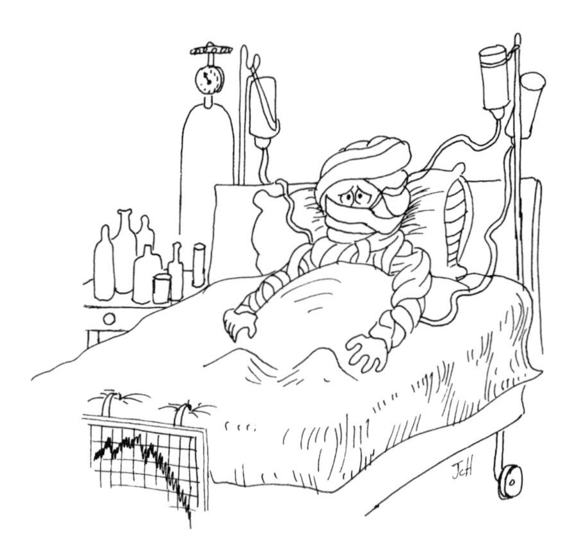

From H. B. Stewart **Grungy George and Sloppy Sally.** *Vantage Press, 1993, p. 48.*

Did you hear about the statistician who drowned while wading in a pond with an average depth of six inches?

Statistics are like a bikini. What they reveal is suggestive, but what they conceal is vital.

The fraction of sick people in hospitals is greater than in the general population. Therefore if you want to increase your chance of being healthy, avoid hospitals.

Residues

A statistician can have his head in an oven and his feet in ice, and he will say that on the average he feels fine.

A statistician is a person who draws a mathematically precise line from an unwarranted assumption to a foregone conclusion.

It's not true that statisticians are a bit weird. Actually they are just your standard normal deviates.

16

SAM SCHWARTZ

Ruler of the Mathematician's Domain[1]

By DONALD E. SIMANEK

Illustrated by JOHN C. HOLDEN

Historians generally dismiss with contempt the notion that mathematicians once dominated a glorious era of human history. We here relate a portion of that dubious history, reconstructed from discrete fragments into a piecewise continuous narrative, which, unfortunately, won't be fully coherent.[2]

The earliest civilization of mathematicians emerged during the late Stochastic Era, when some tribes migrated from the Planes of Euclid, through the Dedekind Cut, to the Kronecker Delta. There they found relatively prime agricultural land, where they settled and multiplied. When locus plagues repeatedly wiped out their lemma groves, they abandoned the area and took to the forests, grubbing for square roots to sustain themselves.[3]

Others migrated to the higher planes, seeking purer air, where they enjoyed the exhilaration from having their heads in the clouds. There they founded the Kingdom of Summaria. Unfortunately this group succumbed to disease—the dreaded Lipschitz Condition—which kills by excruciating attacks of Kurtosis.[4]

---

[1] From *The Vector*, **20** (Fall 1991) pp. 34–38.

[2] The author acknowledges inspiration obtained from the Mathematics Dictionary by James and James (James^2?) They, however, are not to be held in any way responsible for this.

[3] You won't get much out of this unless you have a degree in mathematics. It may help to have a math dictionary handy. Then, again, it may not.

[4] If you've read this far, you are a glutton for *pun*ishment.

In the aftermath of these disasters, the remainder of the mathematicians dispersed into numerous local groups. One of these inhabited the insignificant province of Outer Automorphism. Here was born, of humble origin, a man destined to lead the mathematicians to their greatest glory. His name was Sam Schwartz.

An attack of kurtosis.

Sam's military victories were frequently recounted with awe. In battle he skewed his enemies right and left. He brilliantly confounded the Mesokurtic hordes in the battle of the Konigsburg Bridges. His foes could not determine how to traverse the last bridge without retreating, so, taking advantage of their confusion, Sam forged ahead to capture the Tower of Hanoi. His greatest triumph came when he drove the rebellious Surds from his domain. Eventually all of his enemies' unions were divided, and their forces decimated.

Still, Sam's ambition was unbounded. Through a process of integration by parts, he unified the diverse fractions one by one into an empire. Thus he brought peace to all lands within his compass, which included the area from the Dyadic Trace to the Jordan Curve. Finally, weary of battles, Schwartz turned reflexive, realizing that military conquest was all for nought. He normalized relations with adjoining domains, then turned his attention to peace and domestic affairs. For a while everything seemed affine.

As a ruler, Schwartz was asymptotic to the plight of the poor. He decreed a distributive law to eliminate inequality and ensure increased degrees of freedom. When called upon to adjugate disputes between divergent fractions, he used all of his powers of attenuation to analyze the complex arguments and reach an equitable solution.

A rebellious Surd.

Schwartz became revered as a singular ruler, a statesman of first rank. His royal decrees demonstrated him to be an exponent of rationality. He was widely respected as a man who rigorously held to first principles, and his rectilinearity was without parallel.

At the height of his power, Schwartz built an elegant palace, the Loxodrome. Adjoining it was a university devoted to art, music, and mathematics. This school was constructed on the premises so he could more frequently consult with the faculty. Never content with the least common denominator, Schwartz promoted excellence in all functions of his royal court.

The loxodrome.

In his later years, Schwartz's behavior became singularly eccentric. He lost his powers of differentiation and sometimes even seemed indefinite about integrals. His character was transformed. He took up an odd religion, spending interminable periods seated in the locus position by the light of the Lune, meditating on a point of inflection.

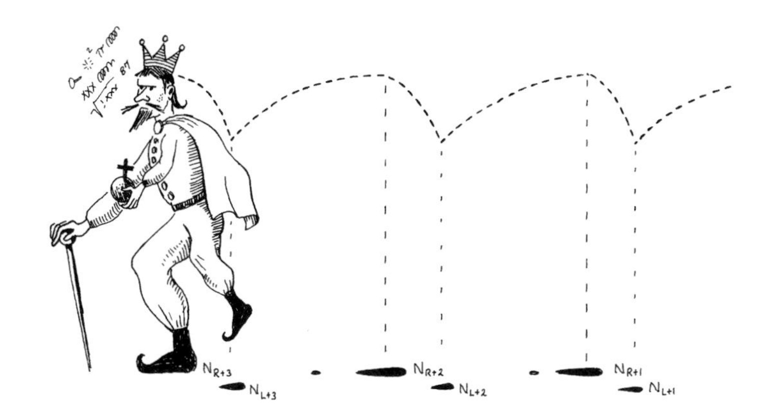

Schwartz became eccentric.

He tolerated numerous deviations and inversions in his court. Rumors circulated that the palace was rife with homomorphisms, and Schwartz himself was suspected of convoluting with a trisectrix. (This was probably untrue, since historians suspect Schwartz was idempotent.)

The palace became a center of isometry and reversion. Some in the court developed an inclination toward mystical philosophies. Persons of high rank frequently attended transcendental functions where they tended to go to extrema. Many were addicted to trivial solutions which led to hyperbolic behavior, and when in their cusps they were capable of only weak convergence.[5]

Degenerate conics.

Even the Church was corrupted by degenerate Conics of deviant declinations.[6] The Church had achieved a position of undue power. Its Cardinal Number had become excessive, and its entire hierarchy was riddled with corruption right down to the level of the local curtate.

To add to these problems, the government bureaucracy had become oblate. Taxes were recursive. A study by the general counting office concluded that the royal treasury's calculations just didn't compute. The finance minister, Pierre Charlier, had decimated the treasury by applying undetermined multipliers to the budget. Checks issued over his signature, Charlier checks, bounced.

Schwartz's speeches became incoherent. He frequently made invalid statements. His decree to realign the province boundaries

[5] If this seems clear to you, please explain it to me.
[6] You might as well give up here; it won't get any better.

by a devious conformal mapping provoked negative reaction. Popular sentiment finally turned against Schwartz when he began to dispense justice arbitrarily. Frustrum and outrage at Schwartz's inequality caused many to call him "Regula Falsi!"

Schwartz's public pronouncements just didn't add up.

Diverse fractions harbored a sense of inequality. Many felt their involute rights were circumscribed. His subjects no longer trusted Schwartz's word. His confidence interval was reduced to the vanishing point.

Schwartz seemed unaware of the magnitude of this coefficient of alienation. He took a lacunary attitude toward rumors of coversed activities against him. He rationalized these as merely an insignificant product of regression among the common factors and vulgar fractions of the population.

The police were determined to circumvent inclinations toward revolution. Citizens suspected of regressive tendencies were taken to the interrogation center to be subjected to third degree equations. Intransitive suspects were tortured on the dreaded tractrix, which could render victims disjoint. The inquisitors were expert at extracting information by the method of exhaustion.

The police had regressed to meanness, and were a force to be reckoned with. Summary trials became the rule, and the prisms were soon filled.

Tortured on the dreaded tractrix.

Even some in Schwartz's own court were outraged at his propositions. A subgroup of the supposedly loyal palace Quaternions formed an axis of revolution to plot Schwartz's downfall. From that point, Schwartz's days in power were numbered.

Their first plot, to drown Schwartz in a Cartesian well, had to be abandoned, for it was the dry season. At a royal banquet for the Italian ambassador, they slipped an extract of limacon tree kernels into Schwartz's riccati. But their calculation of the Poisson ratio was incorrect, and Schwartz only suffered an attack of nausea, which responded to a stiff dose of Euler's formula administered by a parametric.[7]

[7] I never expected you to read this far. Now that you have, I suppose I will have to finish writing the darn thing.

In their frustration they enlisted the aid of the wicked witch of Agnesi to cast a cursoid spell on Schwartz. Her mischievous pet gnomon iterated the numbers of her magic square, causing it to give faulty sums. Its attenuated power merely gave Schwartz a mild case of strophoid fever. After a brief stay in L'Hospital, where he was given homeothetic medicines, he fully recovered.

The wicked witch of Agnesi casts a spell. (A bad number.)

Still, Schwartz was unaware of these radical plots, and of those fomenting a convolution against him. Thus, he was taken by surprise and assassinated by a hired annihilator, while riding his epicycle on his way to a council meeting at the folium.[8]

[8] Threw you a curve with that one!

Sam, riding his epicycle.

Schwartz's funeral was a magnificent function. As he lay in state (at the traditional angle of repose), his soul was commended to the higher powers. Both his friends and his enemies united in wishing that Schwartz would find peace in the Noether World.[9] His mortal residue was encrypted in the royal barycenter, alongside Napier's bones, and a strange bottle containing Klein's ashes.

THE END

Exercises

1. Identify each mathematical term [mis]used in this essay. Give its proper definition. How many are there?

[9] Now aren't you sorry you didn't quit before you began?

2. How many mathematical curves are thrown at you in this essay? Define and sketch each one. Which of these show up in the drawings?
3. How many surds are in the drawing of the "rebellious Surd"?

Endnote

One reader of this little fairy tale complained that I was "picking on a real mathematician." I hope readers do not mistake my fictitious Sam Schwartz for Hermann Amandus Schwartz (1843–1921), who *does* have an inequality named for him. My fictional Pierre Charlier is not to be confused with Carl Wilhelm Ludwig Charlier (1862–1934), who was so often confused by calculations that he invented computational checks which now carry his name (a name long enough to need something to carry it). Charlier's *bank* checks were, so far as I know, good as gold. Mathematician Amalie Noether passed on in 1935, and I have no reason to suppose that her mathematical soul went to a nether world. Anyway, wherever she may or may not be, I am not likely to get any letters of complaint from her, or any of the other mathematicians mentioned in this story.

17
ENGINEERING

To define it rudely but not inaptly, engineering is the art of doing that well with one dollar which any bungler can do with two after a fashion.

Arthur Mellen Wellington (1847–1895) *The Economic Theory of the Location of Railways* (6th. ed., 1900)

Engineering is the art of modelling materials we do not wholly understand, into shapes we cannot precisely analyse so as to withstand forces we cannot properly assess, in such a way that the public has no reason to suspect the extent of our ignorance.

Dr. A. R. Dykes
British Institution of Structural Engineers, 1976

Engineering Hell

An engineer dies and reports to the Pearly Gates. St. Peter checks his list and says, "Ah, I see that you're an engineer. You're in the wrong place."

So the engineer reports to the gates of Hell where he is cordially admitted. But very soon he becomes dissatisfied with the comfort level there, so he sets about designing and building improvements. Before long he's provided Hell with air conditioning, flush toilets and escalators. Naturally this makes him a very popular guy.

One day God calls Satan on the telephone and says with a sneer, "So, how are things down there?" Satan replies, "Just great! You'd never recognize the place now. We've got air conditioning, flush toilets and escalators, and there's no telling what improvements this engineer is going to come up with next."

God replies, "What? You've got an engineer? That's a mistake! He should never have been sent down there. Send him up here right away."

"No way," Satan replies. "We like having an engineer around, so I'm keeping him."

God replies, "Send him back up here or I'll sue!"

Satan laughs uproariously and answers, "Yeah, right. And just where are *you* going to find a lawyer?"

An engineer is someone who is good with figures, but doesn't have the personality of an accountant.

An Arts graduate's view of engineers

Phases of a project

1. Exultation
2. Disenchantment
3. Search for the Guilty
4. Punishment of the Innocent
5. Praise for the Uninvolved

Anon

Electrician's Rule

White's all right.
Green's serene.
Touch the red
And you'll be dead.
Touch the black;
You won't be back.

Safety
Hazards
Often
Can
Kill!

"Should I cut the red wire or the green wire? This is the part I always hate!"

In the USA the color code for electrical wires in houses has the white wire grounded at the main junction box; the green wire is the safety ground, connected to ground separately from the current-carrying wires. The black and red wires are at elevated potential with respect to ground. However, terrorist time-bombs are exempt from such rules.

Handyman's rule: If it moves and it shouldn't, use duct tape. If it doesn't move and it should, use WD-40.

This use of duct tape is not OSHA approved.

Some day it might be possible to tax them.

Michael Faraday (1791–1867) English physicist
(Reply to Gladstone on being asked what was the use of his electrical "toys")

Tell me, Mr Hoover, what are your interests?
Madam, I am an Engineer.
Really? I took you for a gentleman.

Exchange between Herbert Hoover (1874–1964) and a lady he met on a steamship

There are three possible roads to ruin—women, gambling, and technology. The most pleasant is with women, the quickest is with gambling, but the surest is with technology.

Georges Pompidou (1911–74)
French prime minister and president. *Sunday Telegraph* 1968

The theory of electronic devices

Electronic devices are really simple, once you understand the theory of their operation. They operate on smoke. It is very important for each component to contain the correct amount of smoke, which is sealed inside at the factory. If this smoke ever leaks out, the device no longer functions properly.

Illusory gears

A breakthrough in mechanical engineering

The engineering draftsman has long been the unsung hero of mechanical engineering. It is his job to turn an engineer's ill-formed and sometimes half-baked ideas into neat drawings which a machinist can realize as a working mechanism. Yet seldom does the draftsman receive proper recognition. Seldom has a fundamental engineering principle been attributed to the humble draftsman.

Now we can reveal to the scientific world just such a breakthrough. One of the perennial problems of gear design is "backlash" due to the slight looseness of fit where gear teeth mesh. But tighter fits introduce faster wear of the teeth.

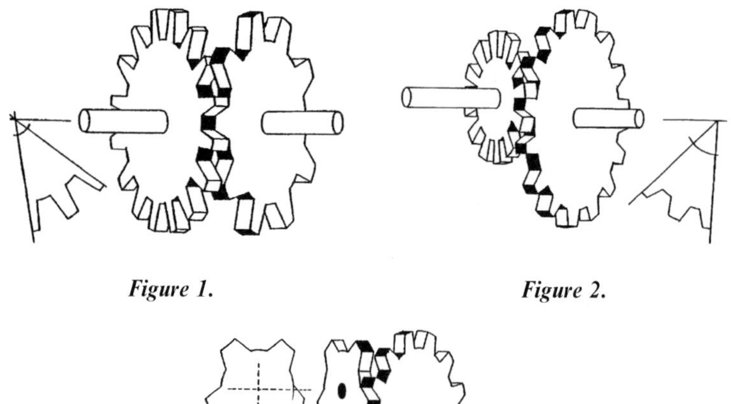

Figure 1. *Figure 2.*

Figure 3.

This dilemma has finally been circumvented thanks to the deceptions inherent in art. Figure 1 shows the basic principle. Perfectly intimate gear mating has been achieved. Look at the gear-tooth where the gears mate. The tooth of the right-hand gear is at the same time the valley *between* two teeth of the left-hand gear. In fact, as gear teeth pass through this area they momentarily vanish, thus greatly reducing friction. In this primitive version of the principle, there do remain some problems with clearance of the gear-teeth.

Figure 2 shows the further evolution of the idea. The left gear has teeth of rather unorthodox shape. Also, we see that the two gear axes lie in different planes. This is a feat hard to achieve with traditional gear design. Still, there remain some clearance problems with the teeth.

The systems shown in Figures 3 and 5 not only solves the clearance problem, but allows two small gears to be driven by one main gear.

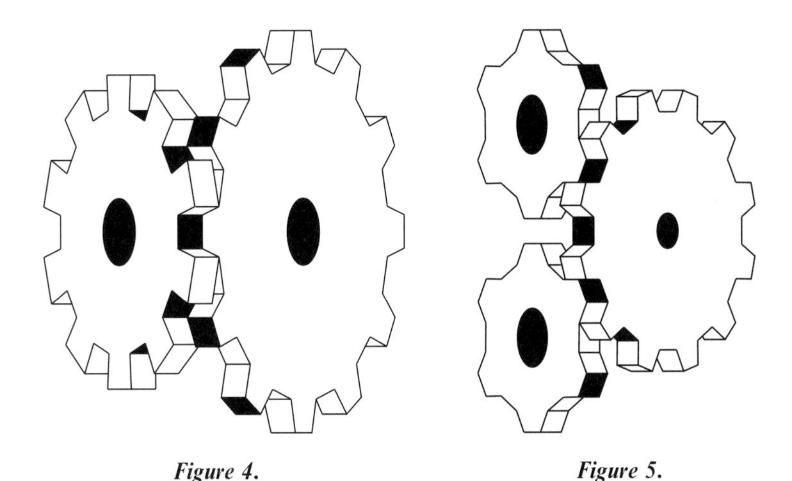

Figure 4. *Figure 5.*

Research continues. Figure 6 shows a prototype sketch of a revolutionary ring gear. It has teeth on the inside and the outside of the ring. But both inside and outside gear teeth are continuous. The perceptive reader will observe that if about 1/3 of either end of the picture is covered, the remainder looks like a normal, old-fashioned gear. Also note that as your gaze moves from left to right (or vice versa) the tops of gear teeth become valleys between teeth (and vice versa). Reverse your scan and they change the other way, but the change doesn't occur at the same place. This subtle shift from tops to valleys equalizes wear on the tooth surfaces when this gear is mated with conventional gears.

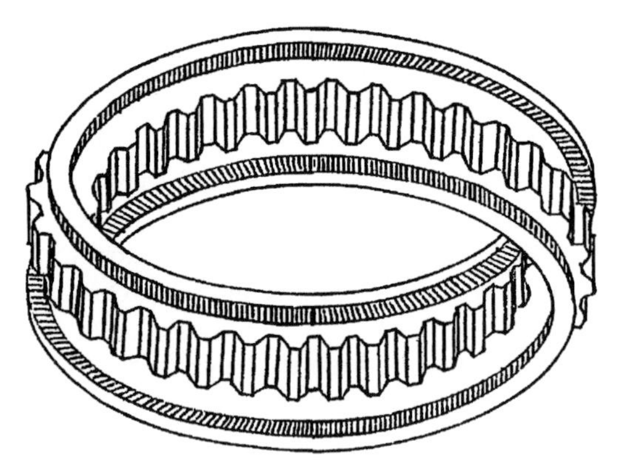

Ring gear.

These gears may truly be called a triumph of art over mundane reality. Research continues to design appropriate gear-cutting tools to mass-produce them.

Anyone desiring to manufacture these gears for use in any mechanism may secure production rights from the inventor. Payment must be made in advance of production.

Comprehending engineers

A matter of priorities

An engineering student parked his bicycle in front of the engineering building. One of his classmates walking by said, "Where did you get that great bike?" The first student replied, "Well, I was walking along yesterday minding my own business when a beautiful girl rode up on this bike. She threw the bike to the ground, took off all her clothes and said, 'Take what you want'." The other nodded approvingly, "Good choice; the clothes probably wouldn't have fit."

Never say die

A priest, a drunkard, and an engineer are being led to the guillotine. The priest is asked whether he wants to lie face up or face down when he meets his fate. The priest says that he would prefer to be facing up so he can look toward Heaven when he dies. They raise the blade of the guillotine and release it. It comes speeding down and suddenly stops just inches from his neck. The authorities take this as divine intervention and release the priest.

Next the drunkard is brought to the guillotine. He also decides to die face up, hoping that he will be as fortunate as the priest. They raise the blade of the guillotine, and release it. It comes speeding down and stops just inches from his neck. So they release the drunkard as well.

The engineer is brought forward. He too chooses to die facing up. They slowly raise the blade of the guillotine, when suddenly the engineer says: "Hey, I see what your problem is."

Professional symbiosis

What is the difference between mechanical engineers and civil engineers? Mechanical engineers build weapons, civil engineers build targets.

The value of experience

An engineer had a reputation of possessing extraordinary ability to diagnose and fix all problems mechanical. After serving his company for over 30 years, he happily retired. Several years later this company contacted him regarding a seemingly impossible problem they were having. One of their multi-million dollar machines wasn't working properly and no one knew how to fix it. The engineer reluctantly took the challenge. He watched as the huge and complex machine was operated, listened carefully to the sounds it made, walked around and observed its functioning from every angle. Then, with a knowing look he asked for a small hammer. Very carefully he located one small part on the machine and gave it one precise tap with the hammer. The machine's raucous sounds changed to a mellow purr, and the company employees verified that it was now operating perfectly.

Soon the company received a bill for $50,000 from the engineer for his services. Thinking this must be a mistake they asked for an itemized bill. The engineer sent one, listing: Tapping with hammer... $1. Knowing where to tap... $49,999.

Failures should be replicable

A software engineer, hardware engineer, and departmental manager were on their way to a meeting in Switzerland. They were driving down a steep mountain road when suddenly the brakes failed. The car careered out of control, bouncing off guard rails until it miraculously ground to a scraping halt along the mountainside. The occupants of the car were unhurt, but they had a problem. They were stuck halfway down the mountain in a car with no brakes.

"I know what we should do." said the manager. "Let's have a meeting, propose a Vision, formulate a Mission Statement, define some Goals, and through a process of Continuous Improvement, find a solution to the Critical Problems and we'll be on our way."

"No," said the hardware engineer. "I've got my Swiss army knife with me. I can strip down the car's braking system, isolate the fault, fix it, and we'll be on our way."

"Wait," said the software engineer. "Before we do anything, shouldn't we push the car back to the top of the mountain and see if it happens again?"

Simulation for the birds

Aircraft and aero engine companies have guns specifically designed to launch dead chickens at the windshields of airliners, military jets and jet engines at maximum velocity. The idea is to simulate the frequent incidents of collisions with airborne fowl to test the strength of the windshields.

An anonymous group of British railway engineers heard about these guns and figured that this method would be perfect for testing the windshields of their new high speed trains. Arrangements were made to borrow one of the guns. When the gun was fired, the engineers stood shocked as the chicken hurtled out of the barrel, crashed into the shatterproof windshield, smashed it to smithereens, crashed through the control console, snapped the engineer's backrest in two and embedded itself in the back wall of the cabin.

The horrified engineers contacted the aircraft engineers, related their experience, and asked what they'd done wrong. The response was just one sentence, "Next time thaw the chicken first!"

Spaced out

During the early years of the space program an astronaut in a spacecraft was being interviewed by radio as he circled the earth. He was asked by a reporter, "How do you feel?"

"How would you feel," the astronaut replied," if you were hundreds of miles above the earth sitting in a vehicle made of 20,000 parts, each one supplied by the lowest bidder?"

During the "space race" of the 1960s, NASA decided it needed a ball point pen which would work in a weightless environment so that astronauts could take notes.

After considerable research and development, the "Astronaut Pen" was developed at a cost of $1 million. The pen worked, and also enjoyed some modest success as a novelty item back here on earth.

The Soviet Union, faced with the same problem, used pencils.

Maybe if they'd talk to each other once in a while . . .

Engineers think that equations approximate the real world.
Physicists think that the real world approximates equations.
Mathematicians are unable to make the connection.
Philosophers ask "What's the real world?"

The dictionary

What engineers say and what they mean by it

Major technological breakthrough. *Back to the drawing board.*

Developed after years of intensive research. *Discovered by accident.*

Project slightly behind original schedule due to unforeseen difficulties. *We were busy working on something else.*

The designs are well within allowable limits. *We just barely made it, by stretching a point or two.*

Customer satisfaction is believed assured. *We are so far behind schedule that the customer was happy to get anything at all.*

Close project co-ordination. *We should have asked someone else; or, let's spread the responsibility for this.*

The design will be finalized in the next reporting period. *We haven't started this job yet, but we've got to say something.*

A number of different approaches are being tried. *We don't know where we're going, but we're moving.*

Test results were extremely gratifying. *It works, and are we ever surprised!*

Extensive effort is being applied toward a fresh approach to the problem. *We just hired three new guys; we'll let them kick it around for a while.*

Preliminary operational tests are inconclusive. *It blew up when we turned on the power.*

The entire concept will have to be abandoned. *The only guy who understood the thing quit.*

Modifications are underway to correct certain minor difficulties. *We threw the whole thing out and are starting from scratch.*

Essentially complete. *Half done.*

Drawing release is lagging. *Not a single drawing exists.*

Risk is high, but acceptable. *With 10 times the budget and 10 times the manpower, we may have a 50/50 chance.*

Serious, but not insurmountable, problems. *It will take a miracle. God should be the program manager.*

Not well defined. *Nobody's thought about it.*

Requires further analysis and management attention. *Totally out of control.*

Glossary of engineering terminology

That's interesting. *I've never seen anything remotely like that before.*

We'll just run diagnostics. *It's a last resort. Maybe it will give us a clue.*

We've noticed some evidence of failure. *Something's burning.*

Major technological breakthrough. *It works only passably well, but it sure looks hi-tech.*

All new design. *No parts are interchangeable with the previous model.*

Rugged. *Too heavy to lift.*

Lightweight. *Lighter than rugged.*

Portable. *Will just fit in the automobile trunk for ease of transport.*

Energy saving feature. *Uses no energy when the power switch is off.*

Low maintenance. *Impossible to fix if broken.*

An extensive report is being prepared on a fresh approach to the problem. *We just hired three fresh university graduates who may think of something we missed.*

We will look into it. *Forget it! We have enough problems for now.*

Please note and initial. *Let's spread the responsibility for this fiasco.*

Give us the benefit of your thinking. *We'll listen to what you have to say as long as it doesn't interfere with what we've already decided to do.*

See me, or let's discuss this. *Come into my office, I'm lonely.*

Years of development. *One of the prototypes finally worked.*

* * * * * *

The student with a Science degree asks, "Why does it work?"

The student with an Engineering degree asks, "How does it work?"

The student with an Accounting degree asks, "How much will it cost?"

The student with an Arts degree asks, "Do you want fries with that?"

To the optimist, the glass is half full. To the pessimist, the glass is half empty. To the engineer, the glass is twice as big as it needs to be.

Doing the impossible

In these days, a man who says a thing cannot be done is quite apt to be interrupted by some idiot doing it.
Elbert Green Hubbard (1856–1915) U.S. author and editor

If at first you don't succeed . . .

Somebody said it couldn't be done
But he, with a grin, replied,
"You shouldn't say it can't be done
At least until you've tried."

So he set to work; armed with a ton
Of zeal he got right to it.
He tackled that thing that couldn't be done
And he couldn't do it.

The dope building

Unknown to most taxpayers, unsuspected even by the *Washington Post*, a sinister and powerful federal agency exists in Washington. It is the Department of Phenomenal Expenditures (DOPE). This agency is responsible for those incredible and unbelievable government projects which the uninformed blame on the "bureaucracy." The DOPE insinuates itself into every expensive federal project (is there any other kind?). DOPE experts evaluate each project, reshaping it into a form best suited for inefficiency, waste, mismanagement and ultimate failure.

Our Washington informant, John Holden, has leaked to us the plans for a new DOPE headquarters building to be located somewhere in the western states. He even supplied a drawing based on those plans.

The DOPE staff designed the entire project, ensuring that it will be fully in harmony with DOPE principles. It will therefore take forever to complete (with an infinite number of cost overruns), at a cost of more money than there is now in the world. Funding will be accomplished by a clever scheme. The project will employ everyone in the country, and each will be taxed on an ungraduated scale of 100%. DOPE historians cite the pyramids of Egypt as a noble precedent for this project.

After an untold number of bribes, payoffs and dirty tricks, the contract for the project was awarded to the firm of Butterfinger, Pfumbler and Goufhoff.

The DOPE building will be powered by a new source of non-synthetically fueled energy: the Robert Schadewald (BS) Gravity Engine.[1] This revolutionary engine, seen in the lower left of the picture, takes advantage of the recently proposed supposition that the universal gravitational constant is not constant, but is decreasing.[2] The engine's massive rotating hammer gains energy from gravity on the down swing. But since the force of gravity has decreased in the meantime, not all of this energy is used in the upswing, so there is leftover energy available during every revolution. This can be used, indeed it must be used, or the engine would speed up continually and tear itself apart.

---

[1] See Chapter 19 for the technical details.

[2] A classic example of a variable constant.

The Dope Building
(from a limited edition print by John C. Holden, 1979.)

Continuing the innovative use of unproven technology, the engine supplies its power through Simanek's Devious Belts. These can link two shafts whose axes are at 90°, in defiance of the laws of mechanics and logic. This invention has been called "shear ingenuity."

The unusual pulley system at the right is known to physicists as the "fool's tackle."[3] There are many kinds (physicists, fools, and tackles). Physicists insist that this is an unworkable pulley system. The DOPE experts don't trust physicists, so those poor fellows on the construction site are given the task of making these tackles work. As you can see, they are valiantly trying.[4]

The building is designed to be earthquake proof. It will prove that earthquakes of 0.5 on the Richter Scale can flatten government buildings. Evidence of the tectonic activity in this region is seen in the volcano in the background, which has already set fire to the paper on which this picture is printed.[5]

DOPE bureaucrats are used to working under strange illusions, and some illusions are incorporated into this building. At the center of the picture we see two workers walking on planks which lie on a level plane, yet they end up one floor below (or above, depending on which way they walk). This subtle principle will be used throughout the building, making elevators unnecessary. Something doesn't seem on the level here, but what can we expect from a government project? On the upper floors we see two engineers sighting to see if everything is aligned. It is.

The ventilation system is another triumph of art over logic. Its ductwork is based on "Schuster's Conundrum" which allows two ducts to feed three without the added cost of joints or elbows.

Though construction has just begun, gaping cracks are already developing in the building. Such cracks have become an integral design feature of federally funded buildings. Past attempts to hide them with shrubbery were not satisfactory, so for this project these are being blended imperceptibly into the landscaping.

---

[3] See Chapter 11 for the technical details.

[4] Physicist readers are invited to calculate the mechanical advantage of this system.

[5] This picture is printed on special illusory paper, as can be seen from the way the corners curl. Hold the page down firmly, for it has a strong tendency to curl up and disappear into another dimension.

18
COMPUTER SCIENCE

Give a man a computer program, and you give him a headache. But teach him to program computers, and you give him the power to create headaches for others the rest of his life.

R. B. Forrest

Computers can figure out all kinds of problems, except the things in the world that just don't add up.

James Magary

The real question is not whether machines think but whether men do.

B. F. Skinner (1904–1990) U.S. psychologist

Computers are useless; they can only give answers.

Pablo Picasso (1881–1973) Spanish artist

It's been my policy to view the Internet not as an 'information highway,' but as an electronic asylum filled with babbling loonies.

Michael (Mike) Royko (1932–1997) U.S. writer, journalist

Laws of computers

1. The most important data will be lost due to parity errors.
2. Unimportant data survives the worst "crash."
3. Two "standard" interfaces are about as similar as two snowflakes.
4. The program that never failed on your last computer will not run on your new computer.
5. Software "bugs" are always infectious.
6. Software upgrades "fix" bugs in earlier versions, while creating more insidious new ones.

* * * * *

We knew that calculating machines were a success when they began to multiply.

Computer expert's glossary

ADA: Something you need to know the name of to be an Expert in Computing. Useful in sentences like, "We had better develop an ADA awareness."

BUG: (1) An elusive creature living in a program that makes its output incorrect. The activity of "debugging," or removing bugs from a program, ends when people get tired of doing it, not when the bugs are removed. (2) That's not a bug; it's a feature.

CACHE: A very expensive part of the memory system of a computer that no one is supposed to know is there.

DESIGN: What you regret not doing later on.

DOCUMENTATION: Instructions translated from Swedish by Japanese for English speaking persons.

HARDWARE: The parts of a computer system that can be kicked.

INFORMATION CENTER: A room staffed by professional computer people whose job it is to tell you why you cannot have the information you require.

INFORMATION PROCESSING: What you call data processing when people are so disgusted with it they won't let it be discussed in their presence.

MACHINE-INDEPENDENT PROGRAM: A program that will not run on any machine.

MEETING: An assembly of computer experts coming together to decide what person or department not represented in the room must solve the problem.

MINICOMPUTER: A computer that can be afforded on the budget of a middle-level manager.

OFFICE AUTOMATION: The use of computers to improve efficiency in the office by removing anyone you would want to talk with over coffee.

ON-LINE: The idea that a human being should always be accessible to a computer.

PASCAL: A programming language named after a man who had nothing to do with it and would turn over in his grave if he knew about it.

PERFORMANCE: A measure of the speed at which a computer system works; or might work under certain circumstances; or was rumored to be working over in Jersey about a month ago.

PRIORITY: A measure of the importance of a user or program. Often expressed as a relative priority, indicating that the user doesn't care when the work is completed so long as he is treated less badly than someone else.

QUALITY CONTROL: Assuring that the quality of a product does not get out of hand and add to the cost of its manufacture or design.

REGRESSION ANALYSIS: Mathematical techniques for trying to understand why things are getting worse.

STRATEGY: A long-range plan whose merit cannot be evaluated until sometime after those creating it have left the organization.

SYSTEMS PROGRAMMER: A person in sandals who has been in the elevator with the senior vice president and is ultimately responsible for a phone call you are to receive from your boss.

How to avoid Y2K problems

Back in 1999 a computer company sent out fax messages to its customers warning them about the Millennium Bug problem.

One customer replied: "Please don't send any more faxes about the Millennium Bug. We do not and will not have any Year 2000 problems as we have not been stupid enough to buy any computers."

Top ten laws of computing

1. When computing, whatever happens, behave as though you meant it to happen, or at least expected it to.
2. When you get to the point where you really understand your computer, it's probably obsolete.
3. When your system is installed and finally working properly, the new model will be available.

4. The first place to look for information is in the section of the manual where you'd least expect to find it.
5. When the going gets tough, upgrade.
6. For every function, there is an equal and opposite malfunction.
7. To err is human. To blame your computer for your mistakes is even more human—it's downright natural.
8. He who laughs last probably has a recent back-up.
9. The number one cause of computer problems is computer solutions.
10. A complex system that doesn't work is invariably found to have evolved from a simpler system that worked just fine.
11. A computer program will always do what you tell it to do, but rarely what you want it to do.

Confidence in incompetence

At a recent software engineering management course, the participants were given an awkward question to answer: "If you had just boarded an airliner and discovered that your team of programmers had been responsible for the flight control software how many of you would disembark immediately?"

Among the ensuing forest of raised hands, only one man sat quietly. When asked what he would do, he replied that he would be quite content to stay on board. With his team's software, he said, the plane was unlikely to even taxi as far as the runway, let alone take off.

Making mistakes more efficiently

An experimental supercomputer was given an extraordinarily lengthy and complex mathematical problem. The researchers eagerly crowded closer to see the answer it produced.

"It's wrong!" one exclaimed. "The computer has made a mistake!"

The astounded experts checked, calculated, and with great disappointment had to admit that the solution was wrong.

Finally the laboratory director said optimistically "Don't be discouraged. Do you realize that it would take 6,300 mathematicians, working 14 hours a day, over 346 years to make a mistake of this magnitude?"

A user-unfriendly computer.

Computer logic

Computers are fundamentally and relentlessly logical, but they have still been found useful in the arts. If Shakespeare had owned a computer, he might have used it this way:

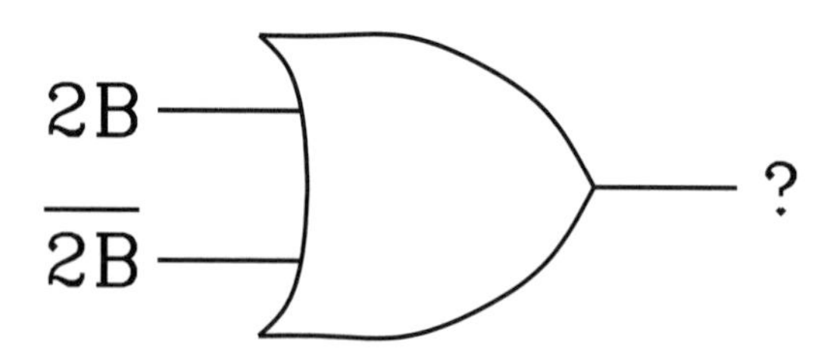

Who says the arts and sciences aren't symbiotic?

In the same vein, here is a flow chart René Descartes might have constructed, had he taken a course in computer programming. [Instead, he wasted his time staying in bed 'till noon.]

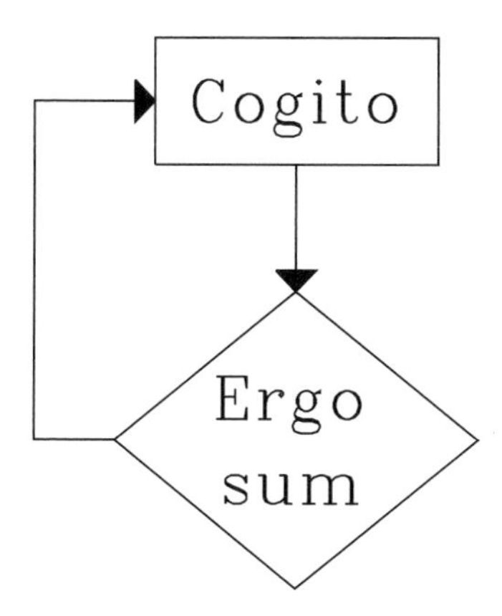

Bytes from the bit-bucket

Computers are great, but can you trust them when the chips are down?

Computers should be forgiving, not forgetting.

Serial connections are sometimes flaky.

Just think, if Edison hadn't invented the electric light we'd all be working at our computers by candlelight.

Computer scientists have developed a computer so fast it can do an infinite loop in 2.5 nanoseconds.

> *I think there is a world market for maybe five computers.*
>
> Thomas Watson (1874–1956)
> U.S. computer industrialist (1943)

If there's artificial intelligence, there's bound to be artificial stupidity.

19
CUTTING-EDGE SCIENCE

If you are on the cutting edge you are holding the knife the wrong way.

DES

There are people who think everything one does with a serious face is sensible.

Georg Christoph Lichtenberg (1742–1799)
German physicist, philosopher, aphorist

Cutting-edge science is a heady combination of inspired speculation, risky extrapolation, leaps of inference, and creative brainstorming. It can lead to revolutionary breakthroughs, or to failure. Scientists working on the cutting edge boldly take a fresh look at existing science and try to push beyond its boundaries. Sometimes it's difficult to distinguish this from pseudoscience and crackpottery.

The essays in this chapter have only one thing in common with genuine cutting-edge science. These take an off-kilter approach to existing science, unfettered by common sense, reality, reason, or good scientific methodology. They are the products of idle minds wondering "What if?" when they ought better to have been doing something constructive.

DISCLAIMER *Any similarity between these essays and real science is purely superficial. Nothing here is to be taken seriously, and anyone seeking research funding for further studies of this kind will be rudely rebuffed and sadly disappointed.*

Vic Stenger's *The Dark Secret of Physics* was used in a science–pseudoscience seminar, presented to students as an example of cutting-edge science. Students were invited to comment critically.

Readers will find other items in this book that may be used in the same manner: *Hazards of Solar Power* (p. 72), and the classic *Ban DHMO* (p. 70). Any who chooses to present these to students as if they were real science is hereby warned that the results may cause you to lose all confidence in the value of education to promote science literacy.

"What goes up" … is basis for a breakthrough

by BOB SCHADEWALD

Gravity! The same mysterious force that keeps our feet on the ground and holds the planets in their orbits was first harnessed centuries ago. The crude wooden water wheels that the Romans built at natural waterfalls have given way to precision turbines at giant man-made dams, but the principle is unchanged. Using water as a conductor, energy is produced from the force of gravity.

Falling water is scarce in many regions of the world. This shortage has led a few visionaries to experiment with engines designed to derive energy directly from gravity, without depending on water or some other intermediary. Many selfless individuals have devoted their lives and fortunes to this dream, only to be derided as cranks. It's true that their so-called "perpetual motion machines" often were complicated contraptions and didn't work very well. Also, many were theoretically unsound.

Because of the past failures in the field of gravitational power, government energy "experts" have written it off, ignored it. While they have poured billions into boondoggle schemes for extracting power from ocean waves, tides, geothermal hot spots and manure, they have lacked the vision to see the potential of gravity.

"If I have seen farther that other men, it is because I've stood on the shoulders of giants," said Isaac Newton. He was referring to men like Descartes and Galileo, who laid the foundations for his theories. Scientific breakthroughs, strokes of genius though they may be, always rest on foundations laid by others.

In recent years, gravitational theorists have completed the groundwork necessary for me to make a major breakthrough.

Many distinguished scientists now believe that the so-called Universal Gravitation Constant is not a constant at all, but is in fact constantly *decreasing*. This crucial discovery means that the acceleration of gravity on earth is constantly decreasing, also. Incredibly, no one seems to have recognized the importance of this concept for gravity engines.

My perfected gravity engine (shown in the diagram) is so simple that you need not be a scientist to understand its operation. A heavy weight of mass m is attached to the periphery of a balanced wheel, and the wheel is started rotating with the weight at the top. As the wheel turns, the weight descends through a distance h, and potential energy equal to mgh is converted to kinetic energy, where g is the average acceleration of gravity during the descent. As rotation continues, the weight again rises to the top, and $mg'h$ kinetic energy is changed to potential energy. As the acceleration of gravity has decreased in the interim, g' is less than g, and there is a net *gain* in kinetic energy. With every revolution, *the wheel speeds up*.

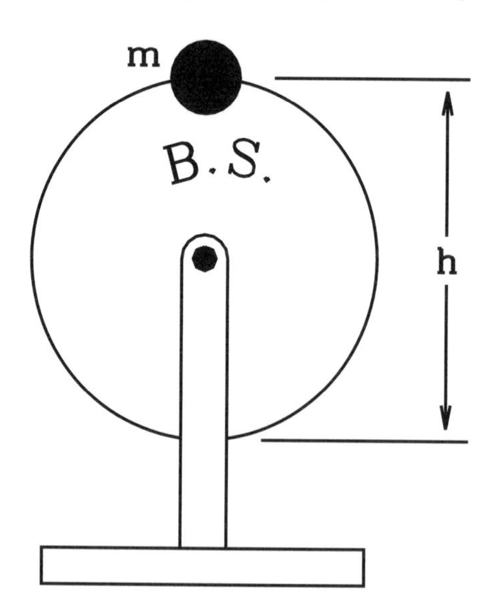

The secret of efficiently extracting energy from gravity is to keep the wheel spinning at a constant speed and siphon off only the excess energy gained on each revolution. The obvious way to accomplish this is to couple the wheel directly to a generator. The excess energy, converted to electricity, then can be fed directly into the existing power grid. No doubt the greatest merit of the Schadewald Gravity Engine is its simplicity—*it has exactly one moving part.* Furthermore, it will run equally well in either direction. These qualities make it especially suitable for use in underdeveloped countries which, incidentally, have been hardest hit by the energy crunch.

The inventor has received absolutely no help—not even encouragement—from any agency, public or private, in developing this boon. He knows that the oil companies, coal companies, the nuclear power industry, and all those holding fat grants for the development of solar power, wind power, and garbage power will band together and try to suppress the invention. Perhaps they will try to have the inventor put away again. Ultimately, the connivers will fail. Though they may destroy the inventor, the invention will live!

It will live to profoundly alter the landscape of the future. Soon you will see row after row of Schadewald Gravity Engines, endlessly producing pollution-free power; hillsides covered with

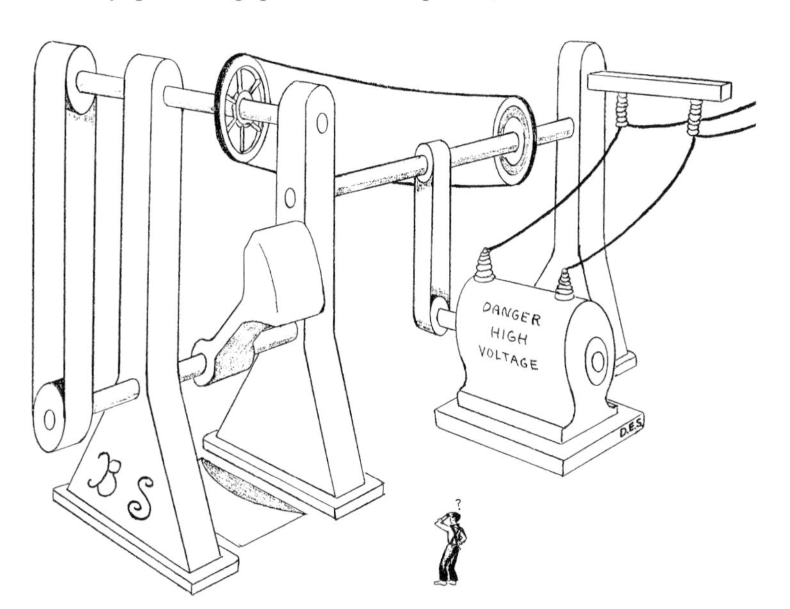

The BS Gravity Engine prototype, driving an electrical generator.

them; mountains, valleys, deserts, even vacant lots, all filled with gravity engines tirelessly pumping out electricity for the benefit of mankind.

That's right; for mankind's benefit, not mine. I admit that, at first, I had visions of grasping the enormous wealth and power at my fingertips, but I prefer to be remembered for my magnanimity rather than my greed.

Therefore, as of April 1st, 1978, I yield my invention to the public domain, that it may solve the energy crisis and bring peace and prosperity to the world. I ask only that my initials be inscribed on the wheel of every engine, so that my genius may get the sort of recognition it deserves.

The dark secret of physics

by TAMERLANE A. EDVARDSSONN[1]

No one can help but notice the conflict between the fundamental physical pictures of the world brought to us by orthodox materialist science, on one hand, and the post-modern sciences on the other. According to the Establishment, we are assemblages of so-called fundamental particles, behaving blindly in a reductionistic universe. In contrast, alternative science brings us a holistic world, with many tiers of reality culminating in an all-pervading Consciousness.

The orthodox view is sustained by raw power, and obfuscating mathematical jargon. But a simple engineering perspective is adequate to expose the pretensions of the high priests of particle physics. We are too easily blinded by the dazzling theory they present us with, while we should be asking for the experimental evidence. Sure, we're **shown** what are alleged to be brilliant confirmations of their Unifications and their Standard Models. But do we think to look beyond the pretty graphs and question **how** this data was obtained?

---

[1] Tamerlane A. Edvardssonn is a longtime researcher into the spiritual sciences, specializing in UFOs and the Interians. He currently lacks a material body, and is channeled by Taner Edis, who passes on his wisdom to the SKEPTIC e-mail discussion group. This document is used here with the permission of Taner Edis.

Few outside the priesthood know what goes on underground in the vast accelerators which supposedly give us the data. So we ignore the mind-boggling complexity of the enterprise. But once you descend into the tunnels and examine the experiments, the scale is shocking. Not only are there complex photomultiplier tubes, but arrays of thousands upon thousands. Layer upon layer of sophisticated detection equipment. Incredibly fast reaction times. Millions of lines of sophisticated computer code which drive the whole system. Uncounted miles of wire. But we should not be impressed! For how can such a complicated, intricately interdependent system work at all? Any halfwit engineer would tell you there are bound to be problems.

By Shaffmacher's Law [1], there is a limit on the reliable operations of complex **mechanical** systems. The error rate grows exponentially with system complexity, so that the output of mechanisms asymptotically approaches sheer randomness with increasing complexity. Careful construction reduces error, but only by modifying the **coefficient** of the exponential, and that with great difficulty. (Note that the same law can be used to show that Artificial Intelligence is impossible [2], hence a non-mechanical **spiritual** principle must underlay human intelligence.)

Both theoretical considerations, and simulation studies under way in the Institute of Higher Noetics [3] indicate that the complexity barrier was probably passed with the earlier cyclotrons, far before experiments of present complexity could be contemplated. Accordingly, only random results should have been produced by these investigations. Why has this not been reported?

There are excellent reasons to suspect a cover-up. Billions of dollars have been poured by the Establishment into orthodox particle physics; a total failure would be a cataclysm way beyond professional embarrassment. Furthermore, such is the confidence of mainstream science in their reductionistic methods that they cannot conceive of gross failure of their models. So they report fake data, fully expecting it to be basically correct, and remediable when the minor glitches in equipment which they think have them stalled are overcome.

Reports confirming this conspiracy have existed for a while, though of course never accessible by normal channels of information. Most recently, a defector from CERN (who has officially been declared insane in order to discredit him) has exposed some of the cover-up, revealing information which corresponds very well to data gathered from previous remote-viewing studies [4].

Normally such a conspiracy would seem to stretch credulity, but recall some peculiar features of particle physics research which makes such manipulation almost trivial. The common paper these days has hundreds of coauthors, each contributing to a narrow aspect of the total experiment. Those who are able to have a global view of the project are only those few leaders who operate more like corporate CEOs than scientists in managing such huge enterprises. They are the ones best able to falsify the results, and also those who would benefit most from such actions. They are also of the class of society who benefits directly from a mechanistic worldview denying human spirit and dignity.

The most important action you can take is to **expose** this cover-up. Spread the word first to fellow investigators in the post-modern sciences. But the world at large must be reached, however difficult their closed mind-sets make the task. Slavery is never pleasurable, and too often getting numb is the only escape.

For more information on the conspiracy, write Institute of Higher Noetics and request **Report #101–A**. To keep up with further developments, join the Institute mailing list, by sending "subscribe" to "pm-physics-request@noetics.edu".

References

[1] Shaffmacher, Robert A. *Journal of Higher Noetics*, 34:6, 5667–5698 (1993).
[2] Erdashov, J. A. *Annals of Postmodern Computing*, 4:1, 1212–1229 (1995).
[3] Escher, Heinz A., & Robert A. Shaffmacher. (In preparation.)
[4] Schadenwold, Karl A. *ParaNatural Philosophy Today*, 13:4, 56–60 (1995).

The age of the universe is a function of time

Scientific estimates of the age of the earth and the universe show a consistent tendency to increase at an increasing rate as time goes on. This relation has been surprisingly consistent during the last three centuries. The implications of this are, of course, profound, for they impact on both the future and the past history of time itself.

I. Introduction

The age of the earth and the universe is of crucial importance to cosmological theories and is of intense popular interest as well. Enough data have accumulated since the early 18th century that we may now critically and objectively attempt to determine whether there's a fundamental underlying relation affecting time itself.

II. Data

1644: Hebrew scholar Dr. John Lightfoot (1602–1675), Vice-Chancellor of Cambridge University, constructed a chronology of history from Biblical genealogies. He calculated that the world was created at the equinox in September of 3298 B.C., at the third hour of the day (9 a.m.). He didn't specify the particular location on earth for which this time applied.

1650: James Ussher (1581–1656), Archbishop of Armagh and Primate of All Ireland, painstakingly correlated Middle Eastern and Mediterranean histories and Holy writ, arriving at the date of creation: Sunday, 23 October, 4004 B.C. No error bars are needed when this date is plotted on the graph, for Ussher considered it exact to the day.

For several centuries thereafter one sees little scientific discussion of the age of the universe, partly because of lack of evidence and theory. People were pondering the question of the age of the earth, and of course, the universe is very likely older than the earth.

1760: Buffon (1707–1788) estimated the earth's age to be 75,000 years by calculating its time of cooling from the molten state.

1831–1833: Charles Lyell (1797–1875) arrived at an age of 240 million years based on fossils of marine mollusks.

1897: William Thomson [Lord Kelvin] (1824–1907) used improved knowledge of heat conduction and radiation to improve the calculation of the earth's cooling rate. Result: an age between 20 and 400 million years.

1901: John Joly (1857–1933) calculated the rate of delivery of salt from rivers to oceans, determining the earth's age to be 90 to 100 million years.

1905–1907: Ernest Rutherford and B. B. Boltwood determined the age of rocks and minerals from measurements of their radioactive decay. They found ages of 500 million years to 1.64 billion years. Subsequent work found rock samples as old as 4.3 billion years.

In the 20th century attention turns from dating the earth, to dating the formation of the solar system, and the universe itself.

1929: Edwin Hubble (1889–1953) interprets the redward shift of distant stars and galaxies as due to expansion of the universe. The rate of this expansion is called the Hubble constant, and if the universe were expanding uniformly since its beginning, this tells us the age of the universe. Extrapolating backward would bring the galaxies together about 2 billion years ago, using Hubble's original figures.

1947: George Gamow (1904–1968) uses Edwin Hubble's (1929) data on luminosity of Cepheid variables to conclude that the universe's "expansion must have started about two or three billion years ago." In a footnote he says, "More recent information leads, however, to an estimate of somewhat longer time periods."

1952: Bart Jan Bok (1906–1983) estimates that the age of galactic clusters must lie between 1 and 10 billion years.

1999: Astronomers working on a special NASA team announced that the universe is about 12 billion years old, based on measurements of the Hubble constant for very distant stars.

III. Analysis

Figure 1 strongly suggests that a real and fundamental relation underlies these data. One may complain that this simplified plot lumps together data on age of the earth and age of the universe. A regression analysis was performed on the two data groups separately. One finds that the earth-age data (before 1929) fit a curve of the form $E = FA^k$ where E is the age estimate and A is the absolute age of the earth at the

AGE OF EARTH

Person	Year CE	Year A (abs)	E = Age Est (yrs)	Method
Lightfoot	1644	5648	3298	Biblical genealogies.
Ussher	1650	5654	4004	Biblical genealogies.
Buffon	1760	5764	75000	Cooling rate of hot earth.
Lyell	1831	5835	240000000	Age of oldest fossils.
Kelvin	1897	5901	210000000	Cooling rate of hot earth.
Joly	1890	5894	100000000	Salt delivery by rivers to oceans.
Boltwood	1907	5911	2.2E+09	Radioactive decay of rocks.
AGE OF UNIVERSE				
Hubble	1929	5933	2E+09	Red-shift of stars and galaxies.
Gamow	1947	5951	2.5E+09	Red-shift of stars and galaxies.
Bok	1952	5956	5E+09	Galactic clusters.
NASA	1999	6003	1.35E+10	Expansion of universe.

Table 1. Estimates of the age of the Earth and the age of the universe.

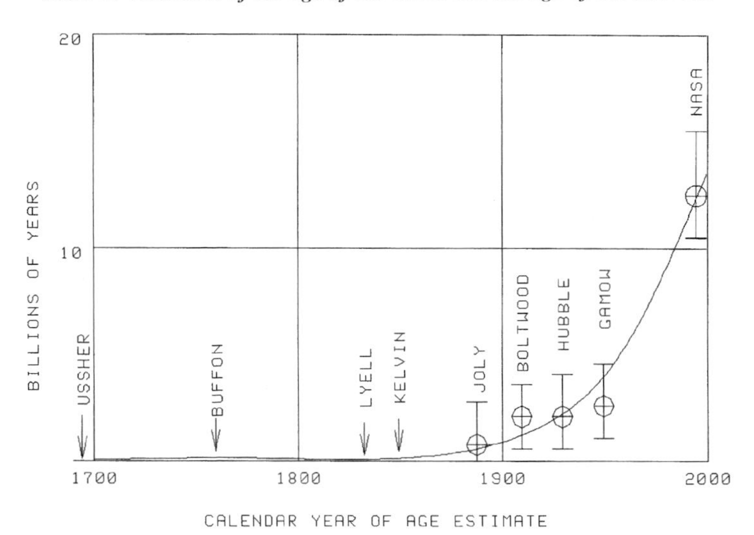

Fig. 1. The estimated age of the universe as a function of the time the estimate was made. Estimates earlier than 1850 are too near the axis to plot, and their error estimates are untrustworthy at best.

time the estimate was made ($A = Y + 4004$) where Y is the calendar year. The fit is best with $F = 3015 \times 10^4$ and $k = 0.0075$. The standard deviation of this fit is about 2.5%.

The estimates of the age of the universe (from 1929 to the present) fit a curve of the same form, $E = FA^k$ with $F = 2910 \times 10^4$ and $k = 0.009$. The standard deviation of the fit is 0.5%. The quality of this fit is much better than that of the earth-age data. This should not be surprising, considering that early estimates of the age of the earth were largely guesswork based on poor methodology, while the more recent estimates of the age of the universe are scientific guesswork based on advanced techniques and instrumentation.

Table 1 gives the raw data, which the reader may use to check these results. These two regression lines extrapolate back to the same point in time, 4004 B.C. This should give encouragement to creationists and other Biblical literalists to stick with that date, and not compromise their beliefs with an earlier date.

IV. Conclusion

Physicists have long treated time as a variable, yet they have failed to realize the full implications of that fact. If these age estimates are to be taken seriously, along with the uncertainties due to methodology, we cannot escape the conclusion that as time goes on the age of universe not only increases, but does so at an accelerating rate.

Since our results are, at present, based entirely on data from the past, this age increase must be happening primarily in the past. Therefore that point in time representing the birth of the universe, the "Big Bang," may be moving backward in time at an ever-increasing rate. The rapidly rising trend of the age curve (Fig. 1) strongly suggests that eventually the universe will be shown to be infinitely old. This will have important consequences:

1. All questions relating to the circumstances in place at the time of the beginning of the universe (the Big Bang) will be relegated to history.
2. All speculation about what was before the Big Bang will likewise be seen as obviously meaningless.

A few theorists suggest alternatives to these conclusions. They say that the trend of age estimates results from a fault of time itself. Perhaps the universal expansion of space is accompanied by a universal expansion of the time frame in which all of this happens. Or, perhaps the space expansion is only an illusion caused by the time expansion. Some say that those theorists just have too much time on their hands.

☞ Fine-print disclaimer: The data and historical facts in the above satire are real, and no bias was used in selecting them, other than the omission of a few obviously wrong age estimates which did not fall on the smooth curve. The regression analysis is, however meaningless. It has been used as an exercise to illustrate some pitfalls of such analysis. The equation has A raised to the approximate power 1/100. This is such a small power that the result is bound to be very near one, for a very wide range of values of A. Looking at it the other way, A is proportional to E^{100}, and therefore the uncertainty in A is 100 times that of E. When extrapolations are made over a large range from a few data points scattered in a very small range, the result has huge uncertainty. The bottom line is that by methods such as these, one can appear to fit almost any data to nearly any preconceived conclusion. Moral for students: do the error analysis properly.

Also, the starting point of the age scale, 4004 was arbitrarily chosen for the analysis, and then used to support the conclusion that the data predicted that date. Compounding the felonies, the very notion of using estimates of age as evidence that the age scale itself is changing with time represents absurdly circular and self-referential logic.

The fact that the conclusions are not even supported by the previous bogus analysis is, in the light of these observations, completely irrelevant to anything.

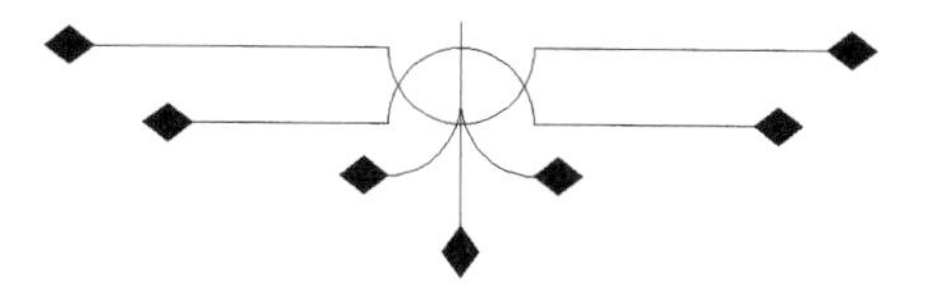

20

RESEARCH

Research is what I'm doing when I don't know what I'm doing.
Werner von Braun (1912–1977) German–U.S. rocket scientist

If we knew what it was we were doing, it would not be called research, would it?
Albert Einstein (1879–1955) Swiss–U.S. physicist

We learn from experience that men never learn anything from experience.
George Bernard Shaw (1856–1950) Irish critic and dramatist

A couple of months in the laboratory can frequently save a couple of hours in the library.
Frank Henry Westheimer (1912–) U.S. Chemist

Researchers have already cast much darkness on this subject and if they continue their investigations we shall soon know nothing at all about it.
Mark Twain (Samuel Langhorne Clemens)
(1835–1910) U.S. Author and humorist

Scientists have come up with a fantastic invention for looking through solid walls. It's called a window.
Richard Feynman (1918–1988) U.S. physicist. Nobel Prize 1965

A most remarkable anomalous event

As everyone knows, when a slice of buttered bread is dropped it invariably lands buttered-side down. This is recognized as one of the fundamental laws of nature.

Yet, on one recent occasion, this law was apparently violated, when a slice dropped by physicist Goufhoff Pfumbler landed butter-side up!

Goufhoff was at first perplexed. Then, struck by a flash of serendipity, he realized that this event might have profound implications. With dreams of publishing in *The Physical Review*, he excitedly shared his experience with a mathematician colleague.

But the mathematician wasn't perturbed at all. "This has a simple explanation," he said. "You buttered the wrong side."

Pfumbler is now planning a research program to determine, *prior to dropping*, which is the "right" side of a piece of bread.

Pfumbler's experiment did reveal another law of nature. He noticed that the hardness of the butter is in direct proportion to the softness of the bread.

From* Science Frontiers *(May–June, 1997).

Omni magazine ran a contest for innovative theories. This attracted the notice of *New Scientist* magazine which reported on the best theories. The winning entry was indeed innovative:

> *When a cat is dropped, it always lands on its feet, and when toast is dropped, it always lands with the buttered side facing down. I propose to strap buttered toast to the back of a cat; the two will hover, spinning inches above the ground. With a giant buttered cat array, a high speed monorail could easily link New York with Chicago.*[1]

There is a deep profundity in this arrangement. S. Voss recognized immediately that a perpetual motion machine had been proposed. He set out to find a flaw. Somehow, energy was being supplied to keep the cat–toast armature turning. Voss observed that any *practical* cat–toast motor would have to be suspended over a very expensive carpet, for the simple reason that the

[1] *New Scientist*, October 19, 1996, p 92.

probability of the toast landing buttered-side down is well known to be proportional to the cost of the carpet. (Linoleum is very poor in this application.) Furthermore, to maintain the machines' efficiency, the rug would have to be frequently cleaned of falling cat-hairs. Carpet cleaning is energy-intensive, and it is here that energy must be supplied, thereby nullifying the perpetual motion claim.[2]

But, on more careful reflection, we note that previous studies of this matter have not properly taken into account modern quantum mechanics and philosophy of science. Like Schrödinger's cat, this poor feline will be caught on its way to the ground state in a mixed quantum state, with equal probabilities of landing one way or the other. Indeed, if this happened in a forest and no observer were there to see it, the outcome would never be decided, the poor kitty remaining in a quantum limbo of half-up and half-down forever.[3]

Other noteworthy theories from the *Omni* contest

(1) A physical explanation of why yawning is contagious: You yawn to equalize the pressure on your eardrums. This pressure change outside your eardrums unbalances other people's ear pressures, so they must yawn to even it out.

Why yawning is contagious.

[2] *New Scientist*, "Feedback", columns for October 19 and November 16, 1996.
[3] *New Scientist*, "Feedback", October 19, 1996, p. 92.

(2) The earth may spin faster on its axis due to deforestation. Just as a figure skater's rate of spin increases when the arms are brought in close to the body. The cutting of tall trees may cause our planet to spin dangerously fast.

Far-out research

In the USA, especially in the midwest, many states have two state-supported universities. Historically one was a "liberal arts" school, the other an "agricultural and mechanical" school. All have grown beyond these historical roots and now offer a full range of majors. But even today students at the liberal arts institution jokingly refer to the other one as "Silo Tech," "Moo U," or "The Udder University."

Mechanical arts has become engineering. Agriculture has become "agricultural science." Serious research is an important component of both. The agriculture department of one university even became involved in the space program. An agriculture professor wondered whether weightlessness could increase milk production of cows. So he proposed sending 50 head of cattle into orbit on a space shuttle. Some thought it might work; others predicted the experiment would be an udder disaster.

Unfortunately, this study was put on hold because of funding cutbacks in the space program. That's too bad, for the experiment had already come to be known as "The herd shot 'round the world."

Rules of the lab

- If an experiment works, something has gone wrong.
- When you don't know what you're doing, do it neatly.
- Experiments must be reproducible; they should fail the same way each time.
- First draw the curves; then plot the data.
- Experience is directly proportional to equipment ruined.
- Always keep a record of your data. It indicates that you have been working.
- To do a lab really well, have your report finished well in advance.
- If you can't arrive at the answer in the usual manner, start at the answer and derive the question.
- In case of doubt, make your account of it sound convincing.
- Do not believe in miracles. Rely on them.
- Team work is essential. It allows you to blame someone else.
- All unmarked beakers should be assumed to contain fast-acting, extremely toxic poisons.
- No experiment is a complete failure. At least it can serve as a negative example.
- Any delicate and expensive piece of glassware will break before any use can be made of it.

Perhaps we need specialized generalists or generalized specialists

A specialist is one who knows more and more about less and less until he knows everything about nothing. A generalist is one who knows a little about everything but not much about anything.

Creative plagiarism

The secret of creativity is knowing how to hide your sources.
Albert Einstein (1879–1955) Swiss–U.S. physicist

I would praise Joad's new book, but modesty forbids.
Bertrand Russell (1872–1970) English philosopher, mathematician and writer

Scraps from beneath the lab bench

It has now been conclusively shown that research is the leading cause of cancer in rats.

Every experiment demonstrates something. If it doesn't demonstrate what you wanted, it demonstrates something else.

The goal of science is to build better mousetraps. The goal of nature is to build smarter mice.

21
LIMERICKS

Classic limericks

There was a young lady named Bright
Whose speed was much faster than light.
She set off one day
In a relative way
And returned on the previous night.

Arthur Reginald Buller (1874–1944) Canadian botanist

The lady was Bright but not bright
And she joined in next day in the flight;
So then two made the date
And then four and then eight,
And her spouse got the hell of a fright.

J. H. Fremlin

To her friends, that Miss Bright used to chatter,
"I have learned something new about matter,
My speed was so great
That it increased my weight;
Yet I failed to become any fatter."

Arthur Reginald Buller

We've heard of that lady named Bright,
And her trip on that fabulous night,
But her increasing mass
Would have soon proved so vast
She'd have been a most singular sight!

It is impossible to travel faster than the speed of light, and certainly not desirable, as one's hat keeps blowing off.

Woody Allen (1935–)
U.S. actor, comedian, director and screenwriter

Once a man in a speedy machine
Passed a light unmistakably green.
All too soon he was dead;
For it really was red,
But the Doppler shift changed what he'd seen.

D. Shirer

Möbius Madness:

A burleycue dancer, a pip
Named Virginia, could peel in a zip;
But she read science fiction
And died of constriction
Attempting a Möbius strip.

Cyril Kornbluth (1923–1958) US writer

Technical note: A Möbius stripper never shows her backside.

A mathematician confided
That a Möbius strip is one-sided.
You'll get quite a laugh
If you cut it in half,
For it stays in one piece when divided.

☞ A model of the Möbius strip may be made from a long strip of paper, giving it a half-twist and gluing the ends together, as shown in the diagram.

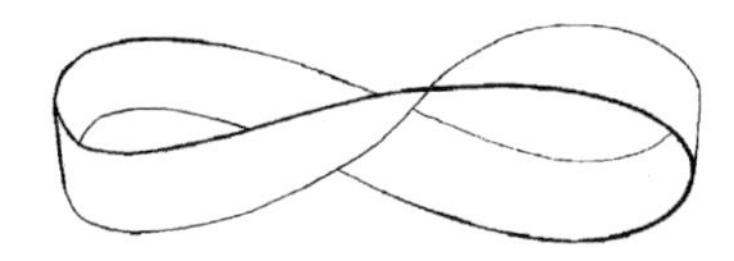

This strip has only one side and one edge. You may verify this by drawing a line down the center of the strip with a pen, moving along the length of the paper and never letting the pen leave the paper surface. Keep going and you will meet the start of your line, which you drew on only one side of the paper. If you examine the strip, you will find there is no other side left unmarked. The paper strip has only one side. You can trace one edge by drawing a line along the edge, and when you reach your starting point, there will be no other edge left to trace.

Now use scissors to cut along the center of the strip, equidistant from the edges. What do you expect will happen when the cut meets its starting point? Try it. When the cut meets its starting point the paper separates into—only one loop, not two! Well, of course it must, for the original strip had only one edge. You never severed that edge, so that edge must remain intact. Your cut created another edge just as long as the first one.

Now make another Möbius strip. Cut it along its length, just $\frac{1}{3}$ of the way from the edge. This cut is much longer than you expect. Predict the result before you complete the cut. Are you surprised?

Of course this can go on to greater heights of topological complexity. Make a strip with a complete twist. Also try a 1.5 twist.

For more about Möbius strips, see chapter 9 of Martin Gardner's *Mathematical Magic Show*, Knopf, 1977. Other books on recreational mathematics discuss them also.

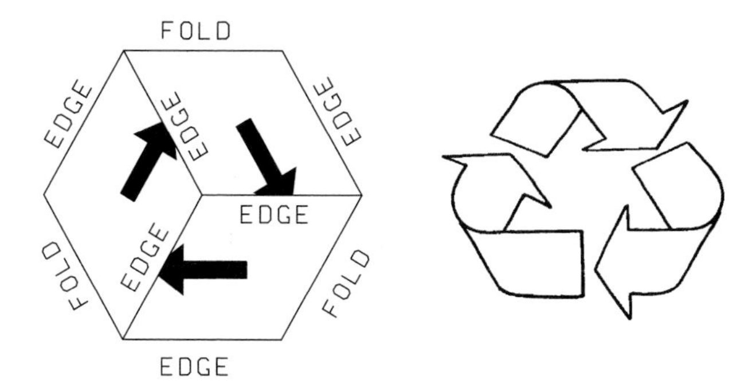

By the way, have you noticed that the recycling icon used in the USA is in the form of a Möbius strip? The shortest Möbius strip you can make of paper would have to lie flat, and look like a hexagon. It's a road-kill Möbius strip.

Scientists often write limericks about their work. Sometimes these are as difficult to understand as their published papers. Unfortunately the authors of the best ones expect credit for their use, so we were reduced to writing some of our own.

Is reality real? Such confusion!
Can it be that it's all just illusion?
Philosophers cogitate;
Scientists speculate.
But none of them reach a conclusion.

You ask, "What's my road-map to fame?
It's gymnastics with math. That's the game!"
If you can't convince them
Then soundly confuse them,
For clear exposition is tame.

Publish or perish! That's the scheme.
We must all play the game "academe."
Winning fame and acclaim;
Is the goal of that game.
It does wonders for one's self-esteem.

Ben Franklin proposed one could fly
A kite to draw charge from the sky,
        Then on down the string
        To a key or a ring.
But do stand inside where it's dry.

A frustrating basic pitfall:
The philosophers' final downfall.
        They try and they try
        But they can't reason why
The real world should make sense at all.

The professor expounds on the world,
Revealing its secrets unfurled
In equations complex
Which thoroughly vex
Any students toward whom they are hurled.

The physicist sooner or later finds
That nature has secrets of subtle kinds.
She does her own thing,
Not helping to bring
Understanding to our finite minds.

The secrets of nature lie hidden.
But some restless minds still are bidden
To learn nature's tricks
With clever physics.
For them not a thing is forbidden.

The wise old professor advises
Those who would seek Nobel Prizes
That picking the locks
On Pandora's box
Is likely to yield up surprises.

The Professor said, "Now I'll tell you
A fact known to only a few
Men or women alive.
Two plus two equals five!
For large enough values of two."

For every action that we can see
An equal reaction there must be.
Their vectors exactly oppose
A fact that conclusively shows
The subtle importance of geometry.

Pronunciation Note:
What did the little acorn
say when he grew up?

Answer: "Gee, I'm a tree!"

Professor Finagle stood aloof.
"That point is not really a goof.
Just a tiny erratum—
That pesky lone datum,
That won't fit my elegant proof."

It's nice to hear wild adulation
After speaking with dry obfuscation
On matters so subtly abstruse
They have no conceivable use—
Lifted fresh from my last publication.

Null vectors have zero projection.
So you ask, "What can be their direction?"
They point any which way.
"That's magic!" you say?
Not really; it's just misdirection.

Professor Finagle's cluttered lab
Is long overdo for full rehab.
His antique equipment
Deserves landfill shipment
And the decor's depressingly drab.

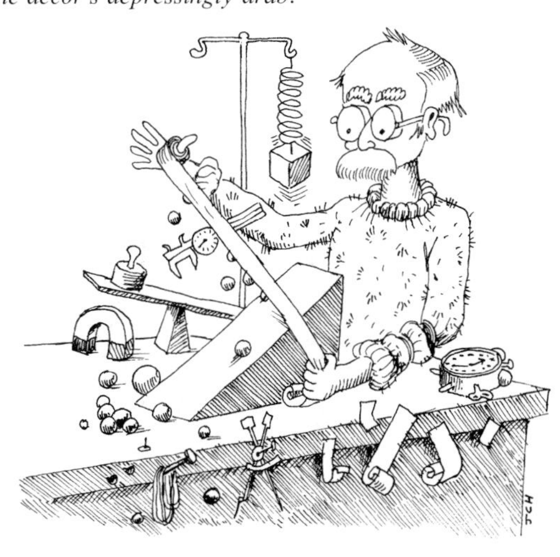

Professor Finagle's lab skills
Are known to be strictly no-frills
He secures everything
With duct-tape and string
And his methods are old as the hills.

A frugal researcher named Dave
Set up his lab in a cave.
His experiments were few
And failed peer review,
But think of the money he saved.

22

Philosophy of science is about as useful to scientists as ornithology is to birds.[1]

Attributed to Richard Feynman (1918–1988)
U.S. Physicist; Nobel Prize 1965

...philosophy is to science as pornography is to sex.[2]

Steve Jones

Philosophers are capable of almost endless enjoyment of mutual misunderstanding.

Lyman Dryson

Philosophy, n. *A route of many roads leading from nowhere to nothing.*

Ambrose Bierce (1842–1914?) U.S. journalist and writer.
The Devil's Dictionary, 1911

There is no statement so absurd that no philosopher will make it.

Cicero, Marcus Tullius (106–43 BCE) Roman statesman.
De Divinatione

Metaphysics is a dark ocean without shores or lighthouse, strewn with many a philosophic wreck.

Immanuel Kant (1724–1804) German philosopher

Metaphysics is a refuge for men who have a strong desire to appear learned and profound but have nothing worth hearing to say. Their

---

[1] Feynman probably adapted it from similar sentiments comparing scholarly studies of a discipline to the actual practice of folks in that discipline. For example, we find "Musicology is to musicians as ornithology is to birds" and "Art criticism is to artists as ornithology is to birds." We were not able to trace this to its source. Feynman's improved wording benefits from the addition of "as useful".

[2] Talking about it is a far cry from doing it.

speculations have helped mankind hardly more than those of the astrologers. What we regard as good in metaphysics is really psychology: the rest is only blah. Ordinarily, it does not even produce good phrases, but is dull and witless. The accumulated body of philosophical speculation is hopelessly self-contradictory. It is not a system at all, but simply a quarreling congeries of systems. The thing that makes philosophers respected is not actually their profundity, but simply their obscurity. They translate vague and dubious ideas into high-sounding words, and their dupes assume, as they assume themselves, that the resulting obfuscation is a contribution to knowledge.

Henry Louis Mencken (1880–1956) *Minority Report, H. L. Mencken's Notebooks*, Knopf, 1956

Philosophy consists very largely of one philosopher arguing that all others are jackasses. He usually proves it, and I should add that he also usually proves that he is one himself.

Henry Louis Mencken. (1880–1956) *Minority Report, H. L. Mencken's Notebooks*, Knopf, 1956

Philosophy is something to think about

Philsophy 101 exercise in thinking.

There are hazards to thinking *too* deeply about something, as this children's poem illustrates.

The centipede.

A centipede was happy quite,
Until a frog in fun
Said, "Pray, which leg comes after which?"
This raised her mind to such a pitch
She lay distracted in the ditch
Considering how to run.

Anon.

Philosophy relies on reason to make sense of things.

There's a mighty big difference between reasons which sound good and good sound reasons.

Henry Wheeler Shaw (pseud: Josh Billings, Uncle Esek)
(1818–1885) U.S. aphorist and humorist

One problem with pre-scientific beliefs in magic and sympathetic influences between things is the problem of distinguishing coincidences from cause-and-effect relations. "Believing is seeing." If you believe something strongly enough, you may see all sorts of confirmations of the belief even when they aren't there.

A philosopher is like a man who goes into a dark coal bin at night to search for a black cat that isn't there.

To which Bertrand Russell added:

A theologian does the same, but always finds the cat.

On the other hand, the genuine causal relations in nature operate the same way every time, even for those who **don't** believe in them.

A physicist had a horseshoe hanging upside down over his office doorway. A colleague said "You don't really believe that old superstition that a horseshoe brings good luck, do you?" The physicist replied, "No, I don't; but I understand it works even if you don't believe in it."

One must never assume that because a hypothesis "works" to account for something in particular that therefore that hypothesis is "true" in general.

A philosophy professor had a bundle of aromatic herbs and spices hanging over the doorway of his office. A colleague asks what it's there for. The professor explains, "It's elephant repellant. And it works, too. Since I've had it hanging there not one elephant has ever dared come in here."

Elephant looking for an open office door.

Scientists know that one should define the outcome of an experiment precisely, and never change the definition of "success" during the experiment or after the experiment.

Two boys were playing in the yard, teaching a dog to do tricks. One said, "He's the smartest dog in the world. Watch this. Bang! You're dead."

The other boy snickered. "He didn't do anything. He's just standing there."

The first boy replied, "That proves how smart he is. He knows he's not dead."

The next day the boys were playing again, and the first said, "My dog knows arithmetic, too."

The other boy was understandably skeptical. "Prove it."

"O.k." He said to the dog, "What's two minus two?" The dog was silent.

"I didn't hear a thing," the first boy observed.

"That's right. He said nothing."

A week later the boys got together again. "My dog is a fast learner. I've got him trained to do more mathematics. Now he can *add* numbers."

The other boy, still skeptical, said, "O.k., prove it."

The first boy said to the dog, "What's two plus one?"

The dog pawed the ground three times.

The second boy said, "Is that all he can do, count to three?"

"Well," said the first boy, "I only said he was trained, not smart."

The ultimate question

A young man, an earnest seeker after truth, had wandered for years, consulting gurus, mystics and wise men in many countries. But still his yearning for ultimate answers was not satisfied. In his travels he had heard of a reclusive guru reputed to be the wisest of the wise, who lived on the top of a Himalayan mountain.

The eager young man outfitted himself to climb the mountain to seek out the wisdom of this guru. After a long, harrowing and arduous climb to the mountain-top the seeker reached the cave of the guru. He was welcomed in, and after a spartan but nourishing meal, the Guru inquired why the man had risked life and limb to come so far.

"I seek answers, oh wise one."

"Then I am sorry to say that you may go away disappointed. I have a strict rule. I answer only one question, no more."

Sitting humbly at the feet of the great Guru, the seeker said: "Then I shall be satisfied with that, for my question is the one ultimate and eternal question, from whose answer all other questions are illuminated."

"Ah; I see that you have thought about this very carefully my son, and I will be pleased to favor you with one answer. What is your question?"

"My question is simply this: Why?"

The Guru raised his eyes to the sky, as if meditating, then looked the seeker straight in the eye and answered with an air of sagacious profundity, "Because!"

Precisely

Some tourists in the Chicago Museum of Natural History are marveling at the dinosaur bones. One of them asks the guard, "Can you tell me how old these bones are?"

The guard replies with an air of authority, "They are 98 million, four years, and six months old."

"That's an awfully exact number," says the surprised tourist. "How do you know their age so precisely?"

The guard answers, "Well, these dinosaur bones were ninety-eight million years old when I started working here, and that was four and a half years ago."

The philosophy exam

A college student was taking his first philosophy examination. On the exam paper was a single line which simply said:

"Is this a question? Discuss."

The student pondered this for a while and then wrote one line: "If that is a question, then this is an answer."

The student received an "A" on the exam.

Beans, philosophy and science

Do an internet search of "beans" and "philosophy" and you'll find a surprising number of references to the pre-Socratic philosophers. This is not to suggest that philosophers are "full of beans," nor is it an oblique reference to the emanations which result from philosophical ruminations. Nor is it an implication that philosophers "don't know beans" about whatever they are talking about.

The Pythagorean sect of the classic Greek era was founded by Pythagoras himself around 500 B.C. Pythagoras is said to have coined the word "philosopher." Pythagoreans wore clothing of animal skins and were strict vegetarians, but were prohibited from eating beans since beans were thought to house the souls of the dead. The bean was also supposed to have arisen early in the creation of the earth and contained the raw materials of which all life is made. One "proof" cited was the fact that if a bean is crushed to a pulp, and allowed to sit in the sunlight for a while, it gives off an odor similar to human sperm. And if you plant a bean in moist earth, as it begins to sprout you may, if you look very closely, see the head of a child growing within it.

Now you should understand that acceptance into the Pythagorean Brotherhood required spending five years listening to the teachings of the Master in total silence, then doing an internship by spending still more time meditating alone in a cave. When the student had achieved the desired mental state, he could actually hear the harmony of the spheres, and maybe even see and hear other wondrous things. He (or she; the Pythagoreans did admit women) might also come to understand the importance of beans in philosophy. On the other hand, students who failed were sent away in disgrace unless they chose to repeat the curriculum by enduring another five years while remaining absolutely silent as more philosophy was stuffed into their minds, followed by another lonely stint in that cave. It's easy to see here the roots of the modern system of higher education.

The sect had other profound ideas, too, such as a prohibition against stirring a fire with an iron poker. And you thought all of the Greek philosophers were rational!

In spite of these symptoms of impaired judgment on the part of the Pythagoreans, they did manage to do some pioneering mathematics, particularly geometry, and applied mathematics to musical scales, believing that the music from perfected and properly tuned instruments was in consonance with the celestial music of the spheres. This has suggested to some that to do mathematics and philosophy it's not necessary to have one's head screwed on straight.

All human beans eventually come to an end.

Now that science has developed ways to gene-splice between parsnips and periwinkles, or beans and humans, what strange things might result?

One of us, John C. Holden, is a closet Pythagorean. He developed and illustrated a whole philosophical system built around beans, including a Supreme Bean. Perhaps the Pythagorean mystics anticipated the Beanie Baby craze of a few years ago.

The Supreme Bean.

The logical necessity of unicorns[3]

A startling discovery was recently made which deserves more attention than it has received: not only do unicorns exist—they could not possibly fail to exist. This discovery is the result of pure logic and was made without the benefit of observation. Since, as everyone knows, our senses are notoriously deceptive, this only makes the

[3] Originally published as "A Recent Discovery which has somehow gone unnoticed," or "My logic is impeccable: don't confuse me with the facts." *The Vector*, **3**, 3, p. 4 (April 1, 1979).

case stronger. The argument proceeds as follows:

U = def: "There is a unicorn in my garden."
E = def: "Unicorns exist."
$U \supset E$ = def: "U implies E," or "If U then E."
$\Box P$ = "P is necessarily true, where P represents any proposition."

1. $U \supset E$
2. $\Box U \supset E$
3. $(\Box U \supset E) \supset \Box E$
4. $\therefore \Box E$

Justification

Premise (1) reads: "If there is a unicorn in my garden then unicorns exist." What could be more reasonable than that?

Premise (2) reads: "If it is necessarily true that there is a unicorn in my garden then unicorns exist." If U is necessarily true then U cannot possibly be false. In other words, if U is necessarily true, then it is true. By premise (1), if U is true, then E is true. If U is necessarily true, then it is true; and if U is true, then E is true. Hence, if U is necessarily true, then E is true.

Premise (3) reads: "If it is necessarily true that there is a unicorn in my garden implies that unicorns exist, then it is necessarily true that unicorns exist." If the antecedent of a true conditional proposition cannot possibly be false then the consequent cannot possibly be false. Or, if the antecedent of a true conditional proposition is necessarily true, then the consequent is necessarily true. (2) is a true conditional proposition with an antecedent which is necessarily true. The consequent, then, is necessarily true. A logician would simply say that what follows from a necessary proposition is itself a necessary proposition. That seems fair.

The conclusion (4) reads: "Therefore it is necessarily true that unicorns exist." (4) follows from (3) and (2) by an eminently sane rule of logic known as Modus Ponens. That rule assures us that when we have "P implies Q" [as in premise (3)] where $P = (\Box U \supset E)$ and $Q = \Box E$ and we also have P [premise (2)], then we are entitled to conclude Q [the conclusion (4)]. Q.E.D.

If some perverted soul wishes to object to premise (3) by maintaining that, if a necessary proposition implies another proposition, it only follows that the other proposition is true, not a necessary truth, the argument may be adjusted as follows:

1. $U \supset E$
2. $\Box U \supset E$
3. $(\Box U \supset E) \supset E$
4. $\therefore E$

—a somewhat weaker conclusion, but nonetheless fascinating.

Howard K. Congdon
Department of Philosophical Zoology

THE UNICORN: Medieval writers describe this magnificent beast as having a mane and reddish-yellow hair, being swift of foot and savage in temperament. It was said that the unicorn would never allow itself to be captured alive, but would die fighting its hunters. However, the unicorn would become docile at the sight of a virgin, and even come to sleep by her side. Someone proposed that a virgin be used as bait to lure unicorns and thereby capture them. It is noteworthy that this method never resulted in the capture of a unicorn.

Scraps of wisdom.

I passed my ethics exam. Of course I cheated.

Did you hear about the fellow who went to the solipsist convention? No one showed up.

Philosophy is a game with objectives and no rules.
Mathematics is a game with rules and no objectives.

Rules for survival at the University:

> Don't LOOK at anything in a physics lab.
> Don't TASTE anything in a chemistry lab.
> Don't SMELL anything in a biology lab.
> Don't TOUCH anything in a medical lab.
> And, most important,
> Don't LISTEN to anything in the philosophy department.

23

KONRAD FINAGLE

Deservedly unheralded giant of science

Libraries are cluttered with biographies of scientists. Rather than retell the familiar tales of well known scientists, it seemed about time someone wrote a totally new biography on a fresh subject. Certainly the world doesn't need yet another biography of Thomas Edison or Albert Einstein.

The aspiring author of biography soon finds that the best known scientists have already been taken. One must find an obscure scientist, someone no previous author considered important enough to write about. Such a scientist turned up quite by accident in the stacks of the Penn State library, where a book was found which had long ago fallen into a dusty corner behind a study carrel. It was written by Konrad Finagle. "Who was *he*?" you may well ask.

Science and engineering students often wonder who invented the Finagle Factor, that indispensable device which enables them to get correct answers even with bad data and invalid methods. Students are well skilled at the process called "finagling" but have no idea who invented it. Now, as a result of years of obscure research in library archives the story of the genius behind this principle may at last be told, adding significantly to the vast amount of misinformation which has been circulated about this question.

Finagle's work was accorded universal neglect during his lifetime, and the essential irrelevance of his work was not fully appreciated until years after his death. Surely he was one of history's most justly unheralded scientists. His work was never equaled. Excelled often, but never equaled.

This brief essay recounts the highlights of Finagle's life and is yet another example of the neglect of history of science in physics

Map of Verlegenstein,
of questionable authenticity.

curricula. After reading this you will surely agree that most of what you have learned about Finagle is wrong.

A note on the illustrations

Nearly all of the illustrations used here are from one of the first widely used physics textbooks, *Ganot's Popular Physics*. Many of its illustrations seem to relate in an uncanny way to the events in Finagle's life. Could it be that some of the illustrations in Ganot's book were borrowed from Finagle's lost autobiography?[1] Maybe not, but they illustrate this story nicely.

onrad Finagle was born in 1858, in the town of Fensterbrechen, located in the tiny European country of Verlegenstein, nestled in the Boravian Alps.

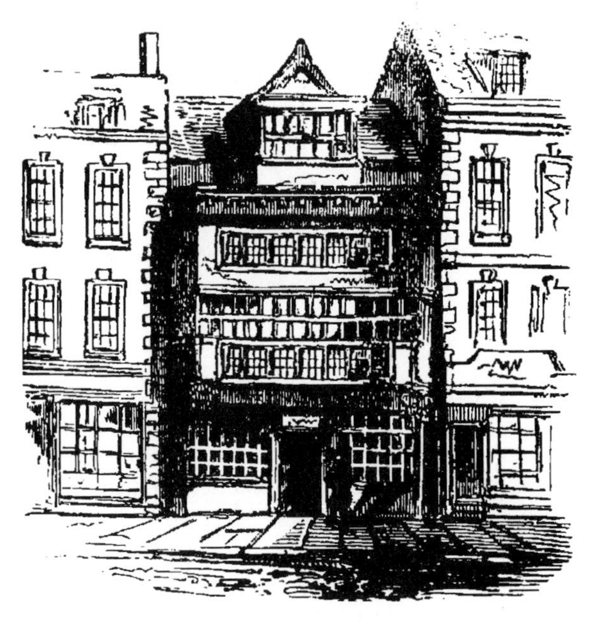

Finagle's birthplace.

[1] This would be especially remarkable since Ganot's book was published in the 1850s, before Finagle was born.

Textbooks of European history totally ignore Verlegenstein. It is so small that few travelers find it. Verlegenstein was created in the 17th century by a slip of the pen while a cartographer was drawing the boundary of adjacent countries. The Verlegensteiners suspected something was amiss when no one collected their taxes several years in a row. So they took advantage of the situation and set up their own government.

Finagle's family was poor but miserable. Konrad's father worked as a clockmaker, his responsibilities including the maintenance and repair of the mechanical clock in the tower of City Hall. Konrad loved to help his father in his work for he was fascinated by clock mechanisms. As he later recalled, this experience taught him an important principle of science: anything can be made right if you tinker with it enough.

Finagle's schoolmaster.

In school Finagle applied this principle to arithmetic. He could always get the right answers to arithmetic problems by adding or multiplying suitably chosen numbers.

His traditionalist and unimaginative schoolmaster was not impressed. Said he, "This boy seems bent upon destroying the whole body of accepted knowledge." This was a curiously prophetic remark. He could not appreciate that Konrad was merely anticipating the new math.

Konrad was only eight years old when his father died tragically in a fall while repairing the tower clock. One of the animated

The Curate.

Young Konrad experiments with the mechanics of rolling bodies.

mechanical figures malfunctioned and knocked him from the tower during the striking of the hour. The local Curate remarked at the funeral, "His time had come."

Konrad and his mother went to live with his uncle Fydor, a wealthy and eccentric tinkerer and inventor. Thus Konrad was suddenly thrust from an environment of poverty and hopelessness to one of wealth and opportunity.

Uncle Fydor kept up with the latest scientific innovations. At parties he often entertained the guests with electrical experiments, such as shocking the ladies with a Leyden jar. He was the life of any party.

Uncle Fydor was the life of the party.

Cousin Eugen ignites his solar cannon.

Cousin Eugen was also a science enthusiast. Finagle was fascinated by his solar-powered cannon.

Finagle uses focused sunlight to ignite paper.

Konrad experimented with even larger solar reflectors. Uncle Fydor was glad to see him taking an interest in science. But he wasn't at all pleased when the library and the back half of the house burned to the ground.

The library burned down.

Finagle uses an electric spark to fire his toy cannon.

In this era the most popular field for amateur experimentation was electricity, so naturally Konrad had to try electrical demonstrations.

An album of pictures from Finagle's childhood

Reflecting on the laws of optics.

Young Finagle takes time out from experiments with rolling objects to study refraction.

Finagle studies resolution of forces and friction.

Finagle was a well-balanced lad with a keen interest in the physics all around him.

Cousin Santislaw tests Finagle's battery made from pickle-jars.

Konrad once made a large battery bank from pickle-jars. Here he has persuaded his cousin Stanislaw to test it. Stan wasn't the brightest member of the family, but he did get a real charge out of this experiment.

Konrad tested another large battery on the family dog, Rover. They buried Rover in the garden.

They buried Rover in the garden.

Surveying class at Unwissenheit Polytechnic.

Konrad's mother wanted to send him to Turin University to study poetry, but Uncle Fydor rejected the idea, saying, "I'll not have my son go to Turin for the verse!" So, instead, Fydor financed Konrad's engineering education at Unwissenheit Polytechnic.

Finagle as a young man.

Upon graduation, Finagle set out, ripe with inexperience, to make his mark on the world of engineering. His first job was as a construction engineer, supervising the construction of twin skyscrapers.

Finagle's first building.

Konrad was now unemployed. To make matters worse, Verlegenstein was hit by the disastrous rutabaga famine of 1884. Finagle decided his future must lie elsewhere. With many others of that era, he chose to emigrate to that fabled land of opportunity, the United States of America.

On arriving in New York City, he found no demand for engineers. The Brooklyn bridge had just been completed, and there was a glut of unemployed engineers.

He chanced by accident on a poster indicating that engineers were needed in Altoona, Pennsylvania. Somehow, with limited funds, he made his way there, only to discover that they wanted people to run locomotives.

Now nearly broke, he was fortunate to find a job at the Pennsylvania State College, the school we now know as The Pennsylvania State University.

At that time Penn State was an agricultural and mechanical school trying to rise from obscurity. The school sought to improve its reputation by upgrading the quality of its faculty. The president was impressed with this young man with his foreign accent and offered Finagle a job as instructor of agricultural mathematics.

Finagle often entertained students with outrageous jokes. He'd tell his math class that famous theorem: "Every horse has six legs—forelegs in front, and two in back." The college president heard of this and asked Finagle why he always told that joke on the *first* day of his Cartesian geometry class. Finagle answered, "Otherwise I'd be putting Descartes *before* the horse." Such humor convinced the president that Finagle was slightly daft, which may be why he promoted Finagle to the rank of Dean.

Finagle liked the Penn State atmosphere, except when the wind blew from the stables. His involvement with agricultural mathematics stimulated several original fertile ideas for research.

This new position allowed Finagle the luxury of a research budget. He set up a research laboratory in the old engineering building in which he spent many hours late at night doing original experiments. Unfortunately we know little of these, for the building was destroyed by the explosion and fire from one of his experiments. The new, modern building which replaced it did *not* carry Finagle's name. In fact Finagle's name is virtually unknown at Penn State today.

The prolific inventor

Finagle was now in the prime of his creativity.

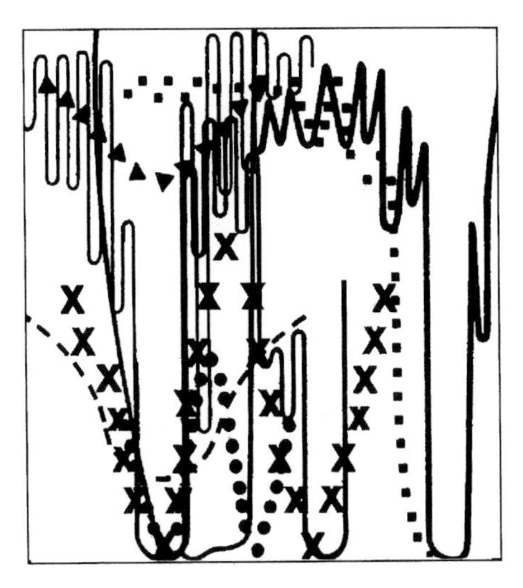

A "wild" graph.

Scientists and engineers must often cope with data which won't fit theory. Finagle introduced several useful methods to deal with such cases—methods still used in engineering today.

1. [Finagle's data exclusion method.] Throw out the data points which don't fit.
2. [Finagle's criterion of acceptability.] If at least 50% of the points fit the curve, the experiment is a success.
3. [Finagle's data enhancement principle.] Scattered data always look more significant when a smooth curve is drawn through the points.
4. [Finagle's principle of significant coincidence.] If two curves on a graph cross, it must be significant.

One day, while passing a store window, Finagle noticed a corset display. Rubber was a relatively new product then. Finagle was impressed at what rubber could do for curves, and suddenly, in a flash of hindsight, he realized this was the solution to the problem of taming "wild" functions.

"Eureka," he shouted, and ran back to campus. He swiped a rubber glove from the biology lab, cut it up, marked a grid on it, and invented *rubber graph paper*. With this tool even the stubbornest data could be brought into line. Some detractors did complain that it could also stretch a point. This corset story, like the story of Newton and the apple, may seem too pat, but Finagle never denied it. When asked if it was true, he always replied, "Of corset is."

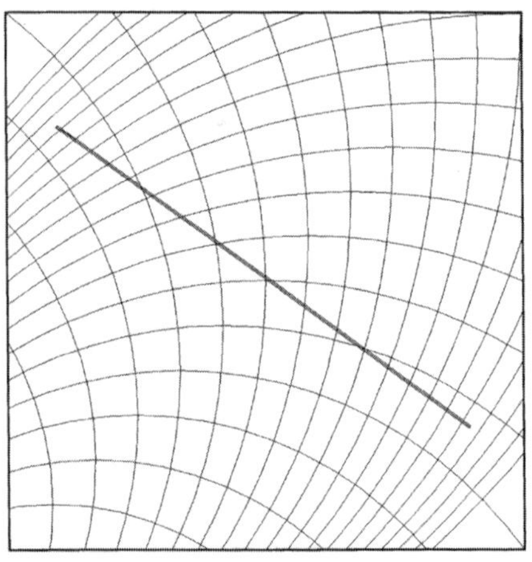

Finagle's rubber graph paper could tame the wildest curves.

Inspiration.

Satirical cartoon: "What next? Rubber slide-rules?". From A. E. Benthic **The Id of the Squid.** ***Compass Publications, 1970, p. 98.***

Finagle's most lasting contribution to science was, of course, his *Finagle factor*. We cannot claim that Finagle was the first to use such "fudge factors," but he probably put them to more imaginative uses than anyone had before.[1]

Finagle justified the use of Finagle factors, saying, "It is futile to do a problem correctly if the answer comes out wrong." But he did advocate subtlety and discretion. In the spirit of Occam's razor, he said, "Never use two Finagle factors when one will suffice."

Scientific ideas are often misinterpreted, misused and corrupted by the general public. Finagle's principles were no exception.

The Penn State football coach appropriated Finagle's principle to his own concerns, saying, "It's not how you play the game, it's whether you win that counts." This principle now pervades all of sports in the truncated form, "Winning is everything."

Sir Eric Dingbat.

That incomprehensible British philosopher, Sir Eric Dingbat, adapted Finagle's rule to philosophy, stating it with uncharacteristic lucidity: "Impeccable logic is of no efficacy if the result is not congruent with the one you know must be correct." And in the worlds of business and politics, Finagle's rules become the essential and universal operating principle: "The ends justify the means."

Finagle never received a Nobel Prize for any of his work, but *other* scientists who used Finagle factors in *their* work received great honor

[1] Things aren't always named for their inventors. The Wheatstone bridge was invented by Hunter Christie, but Wheatstone devised many practical uses for it.

and acclaim. Who can deny that the Finagle factor is the basis of Planck's constant, the Lorentz–Fitzgerald contraction, Fermi's neutrino hypothesis, and many current cosmological theories?

The void theory

In the late 19th century physicists still believed that all of space was filled with a *luminiferous ether*. The ether theory had been formulated by the brilliant French physicist Claude Lumen. He's the one that the unit of light flux is named for. The ether was a subtle medium supposed to fill all of space, providing a medium for light to "wave" in.

Claude Lumen.

Finagle, unable to accept the ether theory, sought an alternative. After ten years of diligent work Finagle proposed his own "Theory of the Void." This theory astonished and confounded mathematicians by its bold use of non-commuting null operators. It was so profound that some claimed that only Finagle and two other persons fully understood it, and the other two had gone insane.

The theory became an immediate sensation. Finagle's popular book on the theory, titled *What's the Void*, was published in 1898. This copy turned up in the Penn State library. It's quite rare. Most libraries threw their copies in the trash long ago. It's worth maybe $2.00 on the rare book market.

One cannot do justice to the Void Theory in this short paper, but here are some highlights, paraphrased from Finagle's book.

Artist's concept of the void (not to scale).

What's The Void?

BY

KONRAD FINAGLE

PROFESSOR OF SCIOSOPHY AND SCIOLISM, THE PENNSYLVANIA STATE UNIVERSITY. EDITOR OF "REVIEWS OF UNCLEAR PHYSICS" AND THE "JOURNAL OF POORLY APPLIED PHYSICS". AUTHOR OF "SHEDDING SOME LIGHT ON LIGHT"; "MATTER OF GRAVITY"; AND "RIGID DYNAMICS OF ELASTIC BODIES."

WITH NUMEROUS EQUATIONS
AND AN ARTISTS' CONCEPTION OF THE VOID

BARNEY NOBEL & CO.
105½ FIFTH AVE, NEW YORK

Title page of Finagle's book.

We are mistaken to think of space as nothing. As Descartes might have said, "If you can think of something, it must exist." And if it exists, we can make theories about it.

Consider what would happen if you took away the space from between matter. Everything in the universe would scrunch together into a volume no larger than a dust speck. We notice that hasn't happened. Why not? Something prevents it. That something is space itself. Space pushes matter apart, opposing gravity. Some matter does manage to slip through space and clump together into atoms, molecules, planets, etc., but it can not collapse completely because *space resists being pushed about*. Space is what keeps everything from happening at the same place.[2]

This is the Finagle principle of the resistentialism of space. It also accounts for reaction forces. One of the most obvious successes of the theory was to explain why a charged particle, such as the electron, does not fly apart from its electrostatic self-repulsion.

Finagle's explanation of Void Theory is a masterful example of reducing a complex theory to its simplistic essence.

[2] Later C. J. Overbeck extended this idea to time: "Time is that great gift of nature which keeps everything from happening at once."

Aristotle said that "Nature abhors a vacuum." In this he was exactly wrong. I say that "The vacuum abhors matter." As evidence, consider how much effort must be exerted by a vacuum pump to fill a bell jar with space.

We must recognize a new conservation law: conservation of space. Space is neither created, destroyed, or filled up. The total amount of space in the universe is constant.

Of course a brief non-mathematical treatment can give only an imperfect description of my theory. The void theory consists entirely of postulates. Having no theorems, there is no chance for faulty logic. All of the postulates are empty of scientific content (as they must be to describe empty space) and have no observable consequences (therefore no experiment can disprove the theory). It is the perfect, self-contained, complete and closed theory.

Finagle's theory was hotly debated. From A. E. Benthic* The Id of the Squid. *Compass Publications, 1970, p. 16.

Finagle's theory shook the scientific world. The void theory was hotly debated at scientific meetings. The popular press had a field day with it. Under the headline "Much Ado About Nothing," the *London Times* ridiculed the void theory, saying, "We suspect that the void exists only in Finagle's head."

Finagle vigorously defended his theory at scientific meetings. At one such conference, attended by the greatest minds of physics, the whole assembly was thrown into a fit of uncontrolled cogitation when Finagle uttered the off-hand remark, "Ether it is or it isn't."

In the ensuing controversy, French scientists backed Lumen's ether, as a matter of national honor. The British and Americans sided with Finagle, and everyone else was befuddled. Even scientists previously sympathetic to Finagle found it hard to accept. Schrödinger was perturbed by it. Heisenberg was uncertain about it. Debye just didn't buy it. And Bohr went into an excited state at the mere mention of it. This is remarkable, for he usually just looked bored.

But Finagle inadvertently killed his own theory. He had earlier written a book analyzing the work of Euclid, and had revealed an insight that others had overlooked, which lay at the heart of Euclid's work and was central to all of mathematics. It is simple to state: "When you can't prove something, make it an axiom."

The young Einstein chanced upon this while he was trying to explain the fact that experiments always gave the same value for the speed of light. Applying Finagle's "axiomatic assertion" principle, Einstein simply *postulated* the speed of light to be constant and derived relativity theory from that. The rest is history. Einstein's relativity soon replaced both the ether and the void theory.

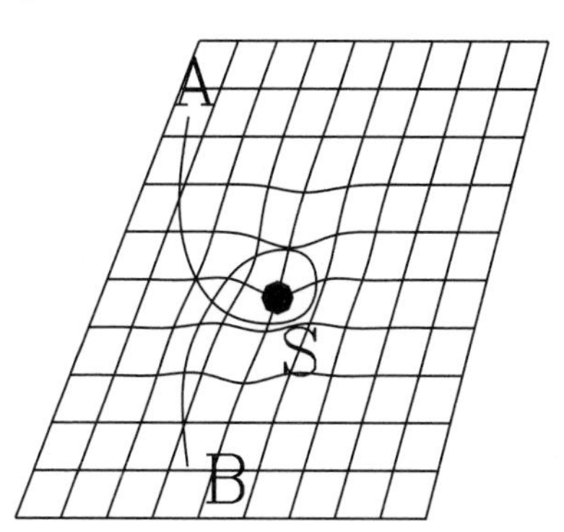

A star (S) warps the metric of space, causing the light path from A to B to curve.

Even Finagle's rubber graph paper stimulated mathematicians to think about rubber-sheet geometry, which led to the interpretation of gravitation as due to warping of the metric of space.

Finagle retired from research and concentrated his energies on teaching. In 1927 he founded the *Journal of Poorly Applied Physics*, and continued as its editor until his death.

Finagle's influence on physics was considerable. His bold void theory *had* changed the attitudes of physicists for all time. It gave them a new way to think about space, and the courage to attribute properties to space. They now talked of "curved" space, and "warped" space. Space had become a valuable scientific resource, flexible and adaptable to the needs of most any theory.

Scientists were liberated from such old-fashioned notions as reality and common sense. It was, in fact, a triumph for Finagle's philosophy: "Results are the only reality; all else is adjustable."

Unfinished business

When Finagle died in 1936 at the age of 78, his fertile mind was still forging new onslaughts on the world of ideas. Two unfinished manuscripts lay on his desk: a text on relativistic statics, and fragments of his definitive work in philosophy of science, *Illogical Positivism*. In this work he summed up his attitude toward measurement:

Finagle in his later years.

> "All of the vexing problems which confront science are the result of measuring things *too* accurately. We must decide whether our object is to create dilemmas, or to get agreeable results."

After Finagle's death many scientists looked back on Finagle's life work and appraised its importance. Albert Michelson, who devoted his entire career to precise measurement, said: "Finagle exemplified all of the things we scientists constantly strive to surmount." DeBroglie observed, "Without Finagle's principles, most of us wouldn't have gotten our degrees." The philosopher Bertrand Russell, commenting on Finagle's place in the history of ideas, said, "Finagle did more to retard the cause of intellectual progress than any other single man." And Albert Einstein, who owed so much to Finagle, acknowledged that, "His contributions to physics were incredible. If he hadn't lived, someone would have had to invent him."

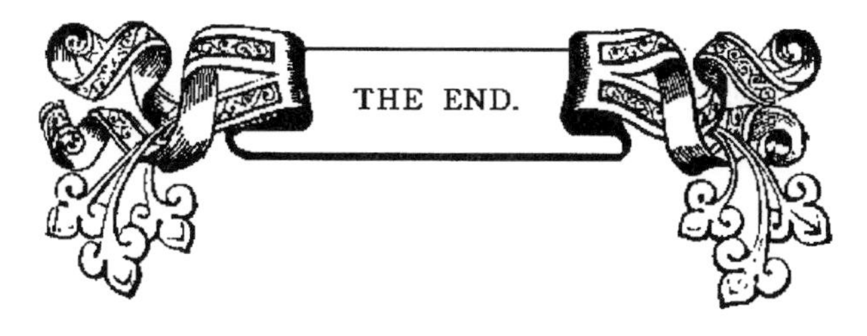

24

SCIENCE AND RELIGION

Bring me into the company of those who seek the truth, and deliver me from those who have found it.

Moses writing the environmental impact statement.

God told Moses he had good news and bad news.

"The good news first," said Moses.

"I'm planning to part the Red Sea to allow you and your people to walk right through and escape from Egypt," said God, adding, "And when the Egyptian soldiers pursue, I'll send the water back on top of them."

"Wonderful," Moses responded, "but what's the bad news?"
"You'll have to write the environmental-impact statement."

* * * * *

Actually it's sacrilegious *not* to accept evolution. Why blame God for everything?

Give a man a fish and he will eat for a day. Teach a man to fish and he will sit in a boat drinking beer all day.

Creative mathematics

James Clerk Maxwell (1831–1879) unified all the phenomena of electricity into four mathematical equations. One mathematical deduction from these equations predicted that electromagnetic fields could propagate energy in the form of waves. Shortly thereafter Heinrich Hertz (1857–1894) demonstrated such waves traveling from a spark gap to a wire coil on opposite sides of his laboratory. We now know these waves as *electromagnetic* waves, which include radio waves, microwaves, infrared, light, ultraviolet, x-rays and gamma rays.

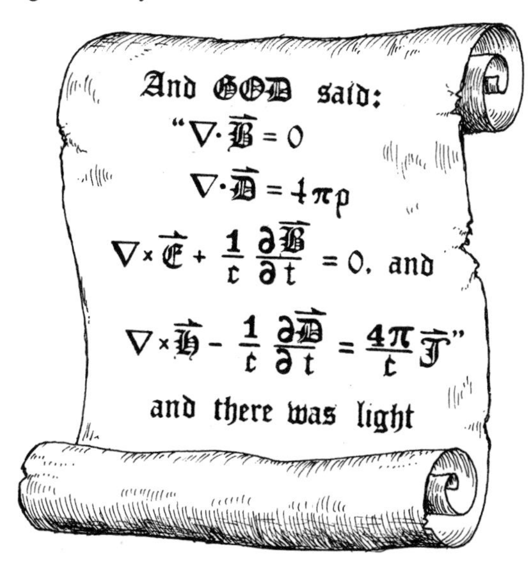

Figure from R. S. Dietz and J. C. Holden* Creation/Evolution Satiricon. *The Bookmaker, 1987, p. 70.

Really tough to get tenure

Academics in American Universities must prove themselves in various ways before they are granted tenure. Tenure conveys the university's approval of the professor's worthiness and carries certain assurances of continued employment. During the probationary period before tenure, the professor's teaching, research, publication record and continuing professional growth are subject to careful scrutiny. When standards are high, it's no wonder that some never receive tenure.

Why God Never Got Tenure

1. God had only one major publication.
2. It was in Aramaic, Hebrew and Greek.
3. It had no references or bibliography.
4. It wasn't published in a refereed journal.
5. Some seriously doubt God wrote it personally.
6. It showed evidence of plagiarism from diverse sources.
7. It may be true that God created the world, but what has God done since then?
8. God's cooperative efforts have been quite limited. There are rumors that he is difficult to work with.
9. The scientific community has had a hard time replicating God's results.
10. God never applied to the Ethics Board for permission to use human subjects.
11. When one experiment went awry, God tried to cover it up by drowning the subjects.
12. When the subjects didn't behave as predicted, God deleted them from the sample.
13. God rarely came to class and just told the students to read The Book.
14. Some say that God had His Son teach certain classes.
15. God expelled his first two students for learning too much.
16. Although He had only ten requirements, most students failed the tests.
17. God's office hours were infrequent and usually held on a mountain top.

A religion for the "rest of us"[1]

By DONALD E. SIMANEK
Illustrated by JOHN C. HOLDEN

It is impossible to imagine the universe run by a wise, just and omnipotent God, but it is quite easy to imagine it run by a board of gods. If such a board actually exists it operates precisely like the board of a corporation that is losing money.

Henry Louis Mencken (1880–1956)
U.S. editor, critic and writer

Many religions make the mistake of assuming that there is only one god, who is all-powerful. That just doesn't square with experience. Look around you at this universe you are in. See what a mess it is? Sure, it "works," after a fashion, but an all-powerful god could certainly have made it better, and simpler to understand.

And look at man, said to be god's greatest creation. Can one imagine any god being proud of such handiwork? Can god do no better at the creation business than to make mankind?

Some religions explain away these obvious facts by inventing an evil supernatural being, a devil, who continually thwarts god's good intentions. But the price paid for this apologetic is to admit that god is not quite all-powerful, or that perhaps god, for inscrutable reasons, permits evil to exist, even though having the power to eliminate it. It has been observed that all existing religions fail when they try to confront the problem of the existence of evil.

It just doesn't wash. The old religious myths mankind invented to explain the existence of everything have failed to account for the world

[1] From *The Vector*, **18** (late 1988), pp. 23–26.

as we know it. It's time for a new mythical account of creation. Fortunately we have devised one which resolves these difficulties completely. We present here its basic tenets, in condensed form.

Designing the Universe.

THE universe was created a long, long time ago (in fact, predating time itself), in a place far, far away (beyond space itself) by a committee of gods, no one knows how many (numbers hadn't yet been invented). The gods were not quite all-powerful, just more powerful than anything else in their neighborhood. Individually they were extremely capable, for they had truly universal minds. But it occurred to some of them that there's little point in having a universal mind if there isn't any universe. So they undertook to create one, since they also had creative minds, and there's little use having a creative mind if you don't create something now and then, if only for practice.

Now each of the gods knew that he (or she, or it—sex hadn't been invented yet) *could* create a perfect universe. Each knew the others couldn't do as well. It soon became apparent that each god had a different conception of perfection.

They worked for cosmic aeons on the design for a universe. Plans were drawn up, argued over, and scrapped. Prototype models were devised, modified, reworked and recalled for defects. Disputes over details became heated. Their deliberations were unruly and chaotic, for *Robert's Rules of Order* hadn't been invented yet. They even had to invent black holes to dispose of their discarded plans and prototypes permanently.

A god-awful row!

Desperate to get the job finished, they formed sub-committees. Each subcommittee assumed responsibility for certain parts of the project, which would later be integrated into the whole. But since each part interacted with every other part, the gods could not agree how to fit them together into a unified universe. So, obviously they never got around to a unified field theory. The most difficult part was man, whom the gods had unwisely decided to make in their own image.

Eventually the disagreements became heated and degenerated into furious arguments followed by a god-awful row of cosmic proportions. This was probably the "big bang" which scientists have called the beginning of the universe as we know it. When the dust settled, all of the gods had been killed. The heavens looked like the

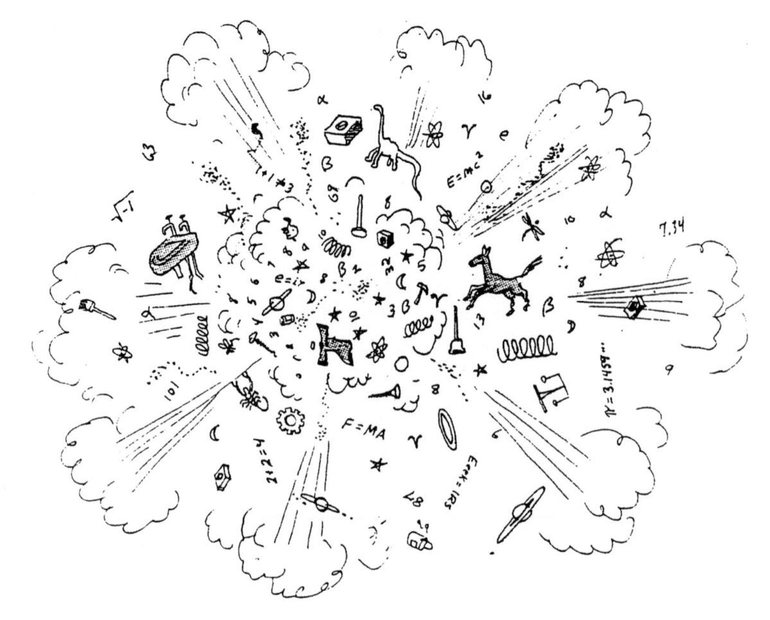

The "big bang."

aftermath of a Shakespearean play. The debris of the battle lay scattered far and wide: time and space, matter and energy, light and darkness, super-strings and sealing-wax, and dust everywhere—all the raw materials which the gods had gathered together to be later assembled into such things as stars, planets and perhaps even intelligent life.

Fortunately the prototype plans for assembly had already been installed in the form of universal laws of physics; for the gods, being lazy, had planned the universe to be self-assembling. Out of this chaos, the raw materials coalesced into some semblance of order.

So we find ourselves in an unfinished, imperfect universe, with no gods left to set things right. We are on our own, to make the best of a bad situation.

You'll notice that some elements of this religion are similar to those found in the more traditional religions. But these universal religious truths were previously imperfectly understood and therefore corrupted.

One unique aspect of this religion is that no god or gods remain, and no heaven or hell. There's no sacred literature, either, for the gods never got a chance to dictate the final committee report.

Yet this religion does a better job of "explaining" the existence and present condition of the universe and the human situation than any other. Having arisen from designs made by the gods of a heavenly committee, we are in many ways an image of them. We see this fact in our human triumphs and failures, brilliant ideas and crackpot notions, cooperative triumphs and divisive conflicts, and of course, in the operation and products of committees.

Is this the one true religion? We make no such claim. Yet it certainly makes more sense than all those other false religions vying for the allegiance of mankind. Why call the others false? Well, they call each other false, and in that appraisal they may all be right!

Creating universes. From R. S. Dietz and J. C. Holden **Creation/Evolution Satiricon.** *The Bookmaker, 1987, p. 16.*

Nor should one *believe* in this religion. That's exactly where most religions go astray. Religions are relatively harmless until you begin to *believe* in them.

However, if you'd care to send money....

The scripturally correct ark

Structural details:

Length = 300 cubits = 450 feet = 138.5 m
Width = 50 cubits = 75 feet = 23 m
Height = 30 cubits = 45 feet = 13.8 m

Three stories, one window, one door, no rudder.

Non-scriptural accouterments:

1. Prow.
2. Roof (necessary to deflect the *very* heavy rain).
3. Wheel house and ladder (to give Noah something to do as Captain).
4. Anchor (for emergencies).
5. Special diet potted plants (eucalyptus tree for Koalas; bamboo for pandas).
6. Tethered ice floes for penguins, polar bears and walruses.

Crew: Noah (Captain), wife (shoveler)
Shem (First Mate), wife (shoveler)
Ham (Second Mate), wife (shoveler)
Japeth (Third Mate), wife (shoveler)

Weather forecast: Cloudy with a 100% chance of very heavy precipitation over the entire earth for 40 days and 40 nights.

Length of cruise: 150 days.

Graphic: Our picture shows the ark shortly after the weather cleared, with the animals happy to be allowed outside for exercise.

Special problems:

1. How did Noah live to be 950 years old without Medicare?
2. What did the animals eat upon release when the cruise was over?
3. When the dove returned to Noah with a "pluct olive leaf," where did she get it?

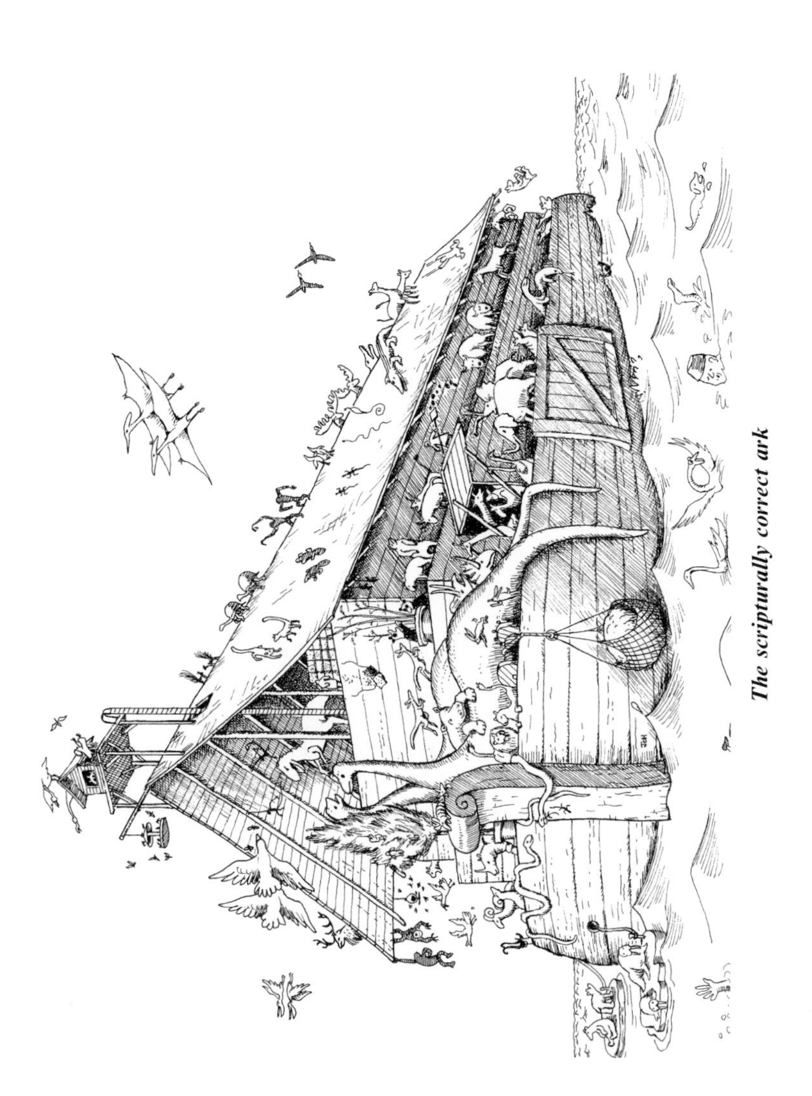

The scripturally correct ark

4. Presumably the added waters from the fountains of the deep and the windows of heaven had salinities of approximately 3.5% to keep ocean creatures alive. Otherwise these animals including whales would have had to be taken on the ark. Did the Ark also contain aquaria for all of the earth's freshwater fish?
5. When did the dinosaurs, ichthyosaurs, mososaurs, pterodactyls and such finally become extinct?

Noah fooling!

Now that the creationists demand equal time for "Bible science" in the schools, particularly the "science" in Genesis, maybe we should discuss such incredible stories as that of Noah's ark with all the seriousness they deserve.

The Bible tells us that Noah's ark was made of "gopher wood." Scholars aren't quite sure what kind of wood that was. Some suspect a mistranslation. Actually the true meaning is easy to guess. As the storm clouds were gathering, Noah and his sons were felling logs for its construction and were behind schedule. Noah had to keep exhorting them, "Go 'fer wood!, Go 'fer wood!"

As the animals were being loaded onto the ark Noah was heard to complain, "Now I herd everything." The animals boarded in pairs, since the singles' cruise hadn't been invented yet.

Noah had to provide specialized food and housing for some animals. Eucalyptus leaves were brought for the Pandas. The bees were kept in the arc-hives.

There wasn't much entertainment on this cruise. Noah couldn't even play cards, for an elephant was standing on the deck. He'd brought along some books to read, but the ark was so crowded that he had to read between the lions.

Noah had to be constantly alert to keep the animals from eating each other. He dared not say, "Let us pray," within hearing of the carnivores.

This story illustrates a problem common today. Noah spent 40 days and nights looking for a parking place, and then had to settle for one way in the boondocks.

When land was finally sighted, everyone was elated. The frogs were hoppy and the pigs went hog-wild. Naturally all of the occupants of the ark were eager to disembark, which raises the stupid question of who got off the boat first. You might suppose it was Noah, but that's not the case. The Bible clearly says: "Noah came forth . . ."

As the animals left the ark, Noah told them to go forth and multiply. After some while, Noah happened upon two snakes sunning themselves. "Why aren't you multiplying?" Noah asked. The snakes replied, "We can't, we're adders."

So Noah had his sons fell some trees in a nearby forest. From these they made a log platform onto which they placed the snakes. You see, even adders can multiply on a log table.

The story of Noah's ark has inspired much speculation, some of which might be described as ark-ane. The ark supposedly contained quite a few animals. While Gen 7:19 says "two of every sort," Gen 7:3 says that fourteen of each of the "clean" beasts were taken (seven pairs, male and female, of all beasts that are ritually clean . . .). Either way, that's a lot of animals to care for and feed, and also a lot of manure to shovel. You have to, too, if you take this story seriously. Remember, there were only eight humans aboard. Since Noah was 600 years old at the time of this trip (Gen 7:11) one wonders how much manual labor he could do. His sons were about 100 years old, according to Gen 5:32. So who did all of the necessary work? Divine intervention might have been required. Ark-angels, perhaps?

Anyway, after the ordeal, it's no wonder that Noah went forth and got drunk (Gen 9:21.)

People sometimes like to pose trivia questions about the Bible, such as: "According to the Bible, who was the first financier?" Answer: Noah, because he floated an entire company when the world was in a state of liquidation. Pharaoh's daughter was a close second, being the first person to draw a prophet from the banks of the Nile.

In the same category is the question: "Where is tennis mentioned in the Bible?" Answer: "When Moses served in Pharaoh's court."

This document was compiled from original obscure and unreliable sources.

The warfare between science and religion

The effort to reconcile science and religion is almost always made, not by theologians, but by scientists unable to shake off altogether the piety absorbed with their mothers' milk. The theologians, with no such dualism addling their wits, are smart enough to see that the two things are implacably and eternally antagonistic, and that any attempt to thrust them into one bag is bound to result in one swallowing the other.

H. L. Mencken (*Minority Report*, 1956)

Andrew White's 1897 book *A History of the Warfare Between Science and Theology in Christendom*[2] attempted a comprehensive account of the hostile relations between science and religion throughout history. White felt that these conflicts were not between science and religion but between science and *dogmatic theology*, and optimistically predicted that religions would abandon dogmatic pronouncements on matters scientific, bringing this warfare to an end. How wrong he was.

Since then we have had over a century of unprecedented scientific advance, and a general (though not total) public acceptance of the beneficial results of science (with very little understanding of the methodology of science). A large segment of the US population rejects those portions of science which conflict with their unbending beliefs arising from a literal interpretation of the King James translation of the Bible. White failed to foresee that.

Biblical literalists particularly dislike evolution, whether it be biological or cosmological. They take the Biblical accounts so literally that they are convinced that the universe was created in seven 24 hour days, the Garden of Eden story was true and factual, and that in this idyllic setting everything was harmonious and perfect. Entropy did not increase, the lion and the lamb were playmates, the speed of light was infinite (or at least very much greater than it is today), and in those early days dinosaurs and man coexisted.

But, after about 1500 years, God became disillusioned with his own creation, causing him to destroy most of mankind with a flood covering the entire land surface of the Earth. This flood was so

[2] D. Appleton and Co., 1897.

The warfare between science and dogmatic theology.

violent that it shaped the Earth as we see it now, including the Grand Canyon, mountains, and the whole lot. The fossils we find now are only the remains of hapless animals which perished in this flood. Only Noah and his family were saved, by building (at God's command) a wooden boat (the ark) large enough to hold all "kinds" of animals on Earth, seven of every clean beast and two each, male and female, of all the others.

After John Lightfoot and Bishop Ussher used Biblical genealogies to determine that the Earth was created in the year 4004 B.C.E., Biblical literalists accepted as gospel (of course) the idea that the Earth is this "young." Modern creationists have relaxed their stand on this age, and are generally willing to accept an Earth age of somewhere between 6000 years and 10,000 years. But they all consider absurd the present scientific conclusion that the Earth is about 4.5 billion years old.

Today a movement called "Creation Science" attempts to present these Biblical stories as a valid scientific theory. Creationists have struggled unsuccessfully for many years to present a solid and credible theory of creation-science. To help them, we have put together some explanations of the subtler points of creation-science, and some visual aids to help readers understand them. This may be one of those rare occasions where a couple of scientists have given creationism the attention and respect it deserves. These illustrations could become standard visual aids in school biology courses if creationism ever is granted "equal time".

We concentrate on cases where Biblical literalism collides head-on with scientific understanding, trampling well established portions of physics, biology, oceanography, and astrophysics. We are confident that the final resolution of these scientific questions will in no way weaken those religious truths which are by their very nature incapable of scientific analysis and verification.

The picture-book of creation-science

From D. U. Wise* Creationism's Geologic Time Scale. *American Scientist (March–April, 1998).

According to creationist Henry Morris (1978), sometime during the 1500 year period between creation week and the flood, Satan's angels and the Archangel Michael fought a great cosmic-scale battle. While tossing massive rocks (asteroids, or whatever solar system debris was handy) at each other, some missed their mark and landed on the Moon. This, Morris says, could account for the Lunar craters.

Since every living thing was created during the first week, and Creationists *do* accept the existence of dinosaurs and other extinct creatures, they suppose that there may well have been dinosaurs in the Garden of Eden. Could the Biblical "serpent" really have been this seemingly friendly dinosaur with evil intent? His intentions may have been stimulated by the realization that Adam and Eve would make a meal far more tasty than the vegetarian diet carnivores were limited to in these perfect times before the fall of man, a time when the world was free of disease, decay and general meanness.

It's not easy to devise a plausible theory to explain where so much water came from during the world-wide flood—enough water to cover even the highest mountaintops. That's a lot of water to account for. Creationists lean toward "And then a miracle occurs!" logic for accounting for such departures from natural laws. Our picture shows how God provides such things. This, of course, does not satisfy scientists.

One creationist theory postulates that the water was previously there in a "vapor canopy" above the atmosphere, and was dumped on the Earth. O.k., but then there's still the problem of where it all

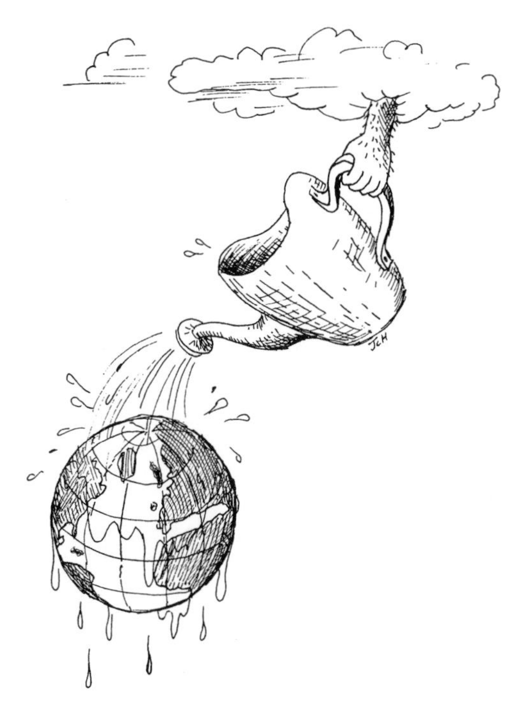

From A. E. Benthic **The Id of the Squid.** *Compass Publications, 86, 2 (March–April, 1998), p. 171.*

went after the flood, and where is it now? We are working on that part.

Creationists have as yet done very little original research. Their only field work has consisted of unsuccessful searches for the remains of Noah's ark. So far these searches have been confined to the Middle East, mostly on Mount Ararat. This may be the wrong mountain! We'd like to help the creationists in this futile search.

Creationists insist that the Genesis Flood was really world-wide, not a local gully-washer. If the earth were entirely covered with water, this body of water would have greatly altered surface currents from those we see on earth now. The mechanisms which drive ocean currents are well-known applications of laws of classical mechanics. On our present earth, ocean currents are constrained to circulate in huge loops (called "gyres") around the ocean's land perimeter. But with no continents sticking up through the surface of this vast ocean there would be nothing for surface currents to "bump up against". Currents on such an earth-covering ocean would extend all the way around the earth in East–West bands, very much like the bands and zones astronomers observe in Jupiter's atmosphere. Such a current is also seen in the West Wind Drift around Antarctica.

The Holy Land is just north of the equator. Therefore, during the global Flood it would lie within a global westward north Equatorial Current. Today's equatorial currents chug along at from 0.2 to 2.0 knots. An ark afloat on such a current for 150 days (Genesis 7:24) would drift 3600 nautical miles to the West and end up in the vicinity of the Atlas Mountains of Morocco in northern Africa. Not only are creationists looking on the wrong mountain; they are looking on the wrong continent!

From R. S. Dietz and J. C. Holden* Creation/Evolution Satiricon. *The Bookmaker, 1987, p. 76.

Those skeptical of the creation-science scenario have difficulty resolving some of its more difficult points. Some animals are found only in remote places like Australia and New Zealand. How did they find their way to the ark's embarkation point if they couldn't swim? How did they get back to their natural homes after the flood? Certain animals became extinct in spite of the fact that God instructed Noah to save them all.

Perhaps there were actually *three* arks. The Bible doesn't say there were not, so we are on solid ground here. Or we may be adrift on troubled waters. One ark contained the dinosaurs, but it sank from overloading. Another wandered off-course and landed in what is now Australia, neatly explaining why certain "kinds" of animals are found only there.

However, just as in real science, one could devise multiple hypotheses for the present-day occurrence of isolated species. The history of science provides several which could easily be resurrected and twisted into forms suitable for integration into creation-science.

Rafting. Some species may have reached their final homes by rafting across the seas on driftwood. Surely lots of driftwood was floating around after the worldwide flood.

Isthmian links. Because of the massive upheaval of the flood, much debris settled out as the waters receded. Some of this debris linked continents with land-bridges or isthmuses which some animals freely walked across. These eventually eroded away.

Island stepping-stones. Likewise, there may have been debris islands after the flood, providing convenient resting places for animals who could swim or fly short distances.

Continental drift. Creationists have not developed a consistent opinion about continental drift. But once that theory is worked out, it may provide the answer. Perhaps before the flood the continents were all contiguous, then drifted apart rapidly afterward as the Earth's surface readjusted from the catastrophic conditions during the flood. After all, the Bible nowhere speaks of more than one continent existing before the flood. Unfortunate pairs of animals may have been separated on different land masses before they had a chance to reproduce, and then became extinct.

The Bible clearly tells us that during the time of Joshua God made the sun stand still for nearly 24 hours. Some creationists have no problem with this since they are also geocentrists who believe the sun actually goes round the Earth. The rest assume that the Sun "stood still in the sky" because the Earth stopped its rotation for nearly a day. This hypothesis presents a few difficulties. Such deceleration and restarting of the Earth would have Earth-shaking consequences far greater than the parting of the Red Sea, and at least as devastating as the plagues of Egypt or even the Noachian flood. These would leave evidences lying around somewhere on Earth, and it's puzzling why they haven't been noticed. Creationist researchers will surely be looking for them as soon as they figure out which mountain the remains of Noah's Ark might be perched upon, and agree upon which old piece of wood they find there is really from the Ark.

Below we see God putting the brakes on the Earth while a heavenly choir sings Kepler's music of the spheres, and the sun looks on in astonishment. Notice, however, that the Earth's motion around the sun has not stopped, as testified by the flow of God's hair (caused either by its initial inertia when God hopped aboard, by the solar wind, or perhaps the motion of the luminiferous ether). Or could it be from the blast of joyous noise from the Heavenly Choir?

For a while creationists cited fossil footprints near the Paluxy river in Texas as evidence that dinosaurs and man lived together. These depressions in the Cretaceous Glen Rose limestone were, they said, footprints made by dinosaurs and man walking the same ancient riverbank. The man-tracks had no toe impressions, an objection easily rationalized away by asserting that these early

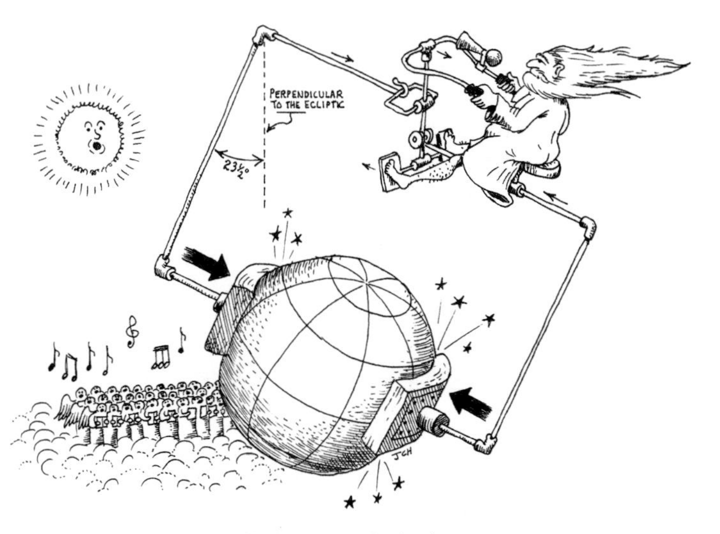

God puts on the brakes.

From R. S. Dietz and J. C. Holden* Creation/Evolution Satiricon. *The Bookmaker, 1987, p. 76.

men must have worn moccasins. The stone surface in which the tracks are found erodes easily from river flooding. During the Great Depression the carving of such fake artifacts became a source of income for locals, resulting in even more detailed footprints. Some pranksters even improved upon the stone depressions in the old river bed. Scientists who have studied the tracks which hadn't been tampered with have judged them to be eroded dinosaur toe tracks; there are no man-tracks. The scenario depicted in our picture is not considered by anyone to be a viable hypothesis.

From R. S. Dietz and J. C. Holden* Creation/Evolution Satiricon. *The Bookmaker, 1987, p. 81.

Creationists say there just was not enough time in their 1500 year scenario for evolution to happen. Well, maybe it happened *very* fast . . .

If the speed of light is really constant then there is a problem with distant stars. Billions and billions of them are so far away that their light could not have reached us in the short time since they were created. Creationists have two ways of dealing with this. (1) In very early times, before the fall of man, everything was perfect, and the speed of light was infinite, or at least very much faster than it is now. (2) God created light coming from stars which aren't there, to give us a false impression of the age of the universe. That explanation seems to suggest that God is deceiving us with evidence of stars which aren't there, and never were there. It also resurrects the old "omphalos" dilemma: "Did Adam have a navel?" If he did that would suggest that God deceived us by giving Adam the appearance of having had a normal birth. And did trees in the Garden of Eden have tree rings? If so, that's evidence of age, and of a history they didn't have. If not, how could they be trees, for growth rings are a characteristic part of the structure of trees? And if they weren't trees as we now know them, how did they evolve (whoops!) from this early proto-tree to trees as we know them? It's a sticky question.

U.S. astrophysicist Thornton Page (1884–1952) made a suggestion which might help the creationists:

> *... what's wrong with creation 5 minutes ago? It would be scarcely more difficult for the Creator to create all of us sitting here, with our memories of events that never really happened, with our worn shoes that were never really new, with spots of soup that were never really spilled on our ties, and so on. Such a beginning is logically possible, but extremely hard to believe!*[3]

Surely any god powerful enough to create the entire universe would also be capable of doing it this way.

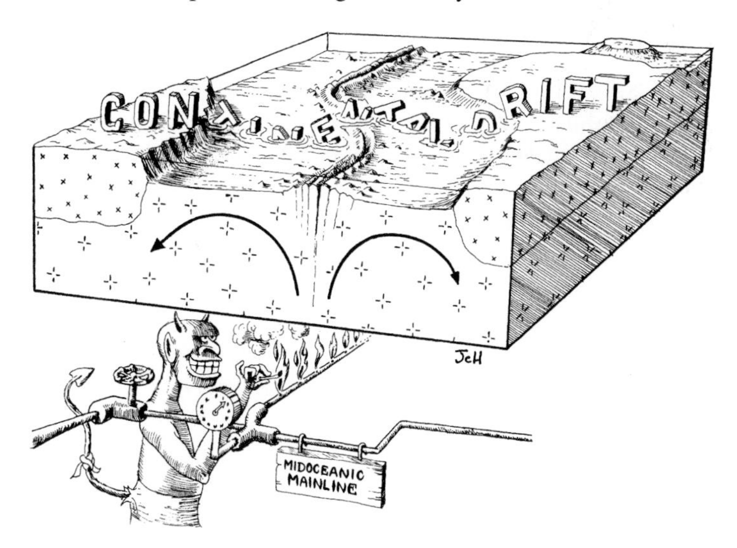

[3] Page, Thornton. *Stars and Galaxies*. Prentice-Hall, 1962.

Any good scientific theory should have explanatory power to help us understand a wide range of phenomena. Most creationists believe in the literal reality of Hell and the Devil. If the Devil's abode really is "the hot place" perhaps this could be a new model for geologists. It could explain the internal heat of the Earth and even, as the picture above illustrates, the reason for continental drift.

To explain how mountains formed, eroded, and formed again in the creationists' short time frame is really quite simple. Since God created time, he had a lot of it left over after the creation. Having so much time on his hands, every so often God does a little light housekeeping, sweeping ocean sediments "under the rug" so to speak.

From R. S. Dietz and J. C. Holden* Creation/Evolution Satiricon. *The Bookmaker, 1987, p. 120.

Some seek a compromise between creation and evolution, but such a thing wouldn't be a marriage made in Heaven.

25
POSTER SESSION

Take PHYSICS.
It will
BLOW YOUR MIND!

If the gods
had meant us to understand

PHYSICS

they'd have made it
simpler!

Some people are afraid of physics. Wait until they find out it's all around them!

Time is that great gift of nature which keeps everything from happening at once.

— C. J. Overbeck

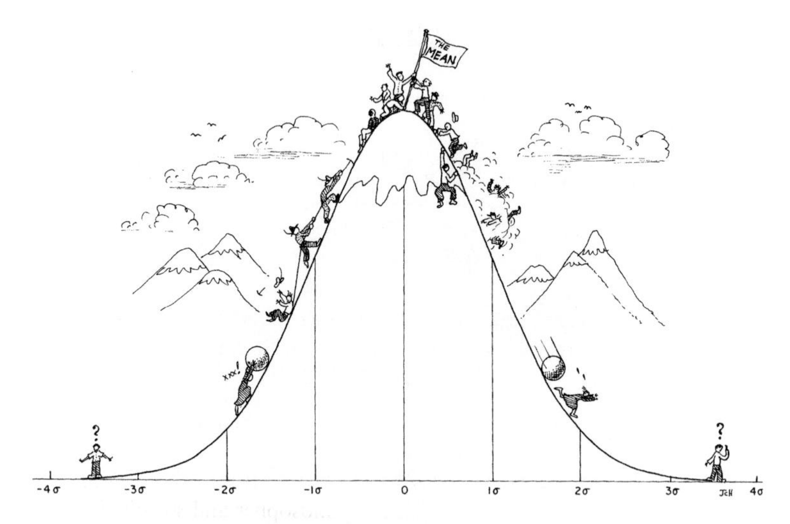

Being average is a mean accomplishment.

26
FINAL EXAM

Examinations are formidable even to the best prepared, for the greatest fool may ask more than the wisest man may answer.
C. C. Colton (1780–1832) English author

When a man's knowledge is not in order, the more of it he has, the greater will be his confusion.
Herbert Spencer (1820–1903)
English philosopher and sociologist

Albert encounters his first true–false exam.

1. Which of the following is *not* an answer to this question?
   (a) All of the responses below.
   (b) This one.
   (c) Some of the above.
   (d) All of the above.
   (e) None of these.
2. Give the name of the substance which will always obey Boyle's and Charles' laws exactly and under all conditions.
3. (a) What is the top speed of a garden snail moving across the

ground in search of juicy leaves to munch on? (b) Express this snail's pace in centimeters per second. (c) Express this also in the units furlongs/fortnight.

4. Why is it that lightning never strikes twice in the same place?
5. What animal needs the least nourishment?
6. Why do balloons rise in the atmosphere?
7. How does the man in the moon style his hair?
8. Pound-for-pound (or kilogram-for-kilogram if you like), what is the most vicious animal on earth?
9. What has two arms and a neck but no head; and where its feet should be it has a tail instead?
10. What pine has the longest and sharpest needles?
11. Why does the giraffe have such a long neck?
12. Seven is an odd number. How do you make it even?
13. Define cardinal numbers.
14. Why does heat travel faster than cold?
15. Why are summer days longer than winter days?

Answers

1. I haven't the foggiest idea.
2. Bisnordeoxydehydroethyl alcohol
3. The garden snail is the speediest land snail, with a speed of 0.03 miles/hr. An internet source gives it as 0.0313, but all these figures are surely not significant. A biologist estimated 0.05 cm/sec. A student consulted the Guinness Book of World Records which gave a value of 0.092 in/sec. Another student says the same source gives 55 yards/hr; that's 0.03 mi/hr. A biology book gives 2 inches/minute. Obviously these sources weren't observing the same snail. So much for "authoritative" sources. We'll go with 0.03 miles/hr.
   1 furlong = 1/8 mile = 220 yard.
   1 fortnight = 14 days = 2 weeks.
   (0.03 miles/hr)(8 furlong/mile)(24 hr/day)(1/14 day/fortnight) = 0.41 furlong/fortnight.
   Some people put out beer in their gardens to discourage snails and slugs. I'm not certain whether it kills them, or just distracts them from eating the plants. We won't even consider here how fast a drunken snail can crawl.

4. Because after it hits once, the same place isn't there any more. Alternate acceptable answer: That place isn't the same place any more.
5. The moth. It eats nothing but holes.
6. Because someone let go of the string.
7. 'E clips it.
8. The amoeba.
9. A shirt.
10. The porcupine.
11. Because its head is so far from its body. And it has such long legs because its body is so far above the ground.
12. Erase the "s".
13. They are what you use to count crested red birds.
14. Because you can catch cold.
15. Because heat expands them.

This exam is for the birds

An ornithology professor devised a novel final examination. When the students arrived to take the exam they saw on the lecture desk several dozen numbered boxes, each containing a bird. The boxes had holes in the bottom for the bird's feet to stick out. The professor explained that the students were to identify each bird by examining its feet.

From J. C. Holden et al.* Now That You Are Here... A Guide to the Methow Valley. *The Bookmaker, 1976, p. 96.

Most of the students took the exam very seriously, carefully examining the feet protruding from each box. Some shook the boxes when no one was looking, hoping to get some response from the bird to help in the identification. But one student became more and more frustrated and angry. When the time was up, this student confronted the professor and strongly stated his opinion that the exam was one of the dumbest ideas he'd ever heard of, that it was unreasonable to test students in this way, that the entire course was worthless and that the professor was incompetent to teach the subject.

The professor was taken aback by such audacity on the part of a student, but he kept calm. "I'm sorry you feel so strongly negative about this. But I will take your opinions under advisement." He got out a pencil and asked, "What is your name?"

The student held out one leg to the professor and said, "Read my feet!"

This is actually a cleaned-up version of the story. In the original, only the birds' tails protruded from the boxes.

* * * * *

End-of-term lament: "I can't think, therefore I cram."

27

The Illustrated Dictionary of Physics

An introductory physics course introduces more new technical terms than a foreign language course. Pity the poor student who must learn these. Pity the poor teacher who must juggle these with precision.

Words for technical terms are often borrowed from common language, but the words take on new and precise meanings. The physical meaning of "work" is not necessarily what you get paid for. The student who puts in "a lot of work" studying for an exam actually does very few joules of work. And "moment of inertia" in physics does not mean those periods when students nap in class.

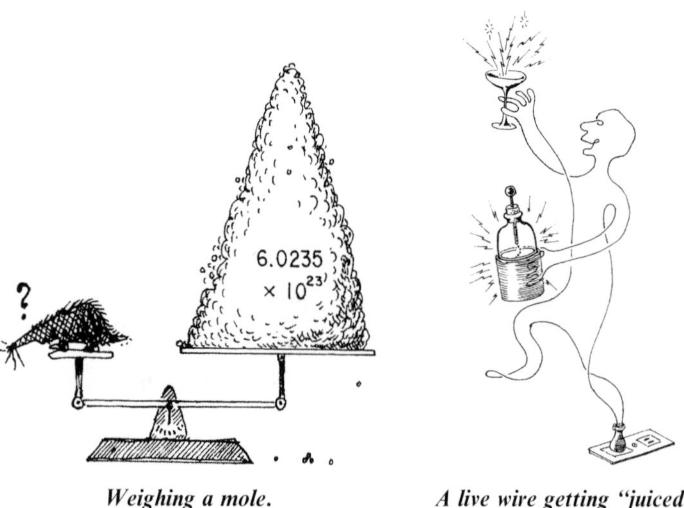

Weighing a mole. ***A live wire getting "juiced".***

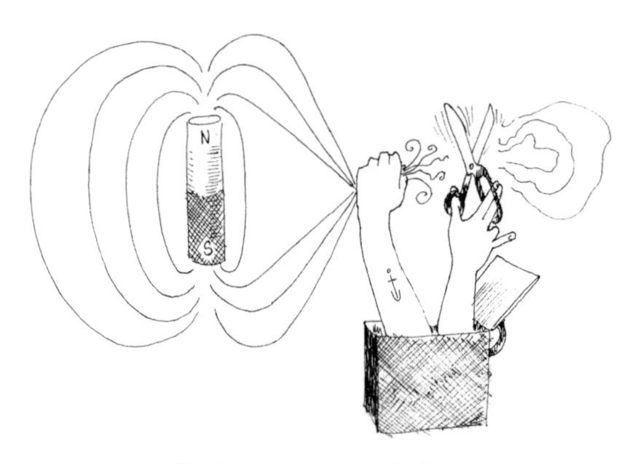

Cutting magnetic field lines.

Chemists talk about moles and weighing moles. Physicists speak of charge-carrying wires and cutting field lines. What if they really meant what those names imply? John Holden helps us visualize that.

Units and dimensions are constant bugaboos for physics students. They are used to distinguish numbers which represent measurements of different things. Students learn that one mustn't add numbers with different units, such as length to mass. That would be akin to adding apples and oranges, which would, however, be o.k. if one only wanted to count pieces of fruit. One may multiply different units, and the results are in still different units. Work is the product of distance and force. If these have units of feet and pounds, the work has the unit "foot-pound." But a quite different concept, torque, is also the product of a distance (lever arm in feet) and a force (in pounds). This, however is a vector, while work is a scalar. To distinguish one from the other, and keep everything perfectly clear, torque is given the unit "pound-foot."

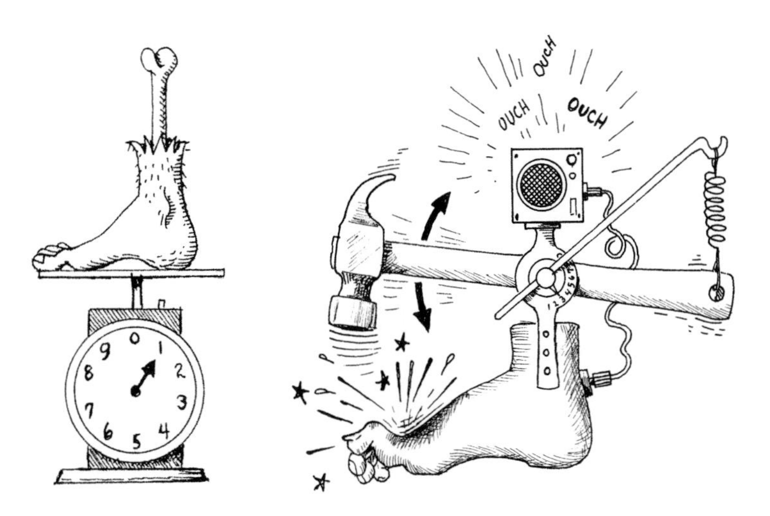

The pound-foot and the foot-pound.

To make life more interesting, there's also the foot-candle, a unit of luminous intensity.

So far we've mentioned only units of the older "English system," used nowadays only in backward countries such as the USA. Other quaint units of this system are the slug and the poundal, both units of mass, and the horsepower, a unit of work.

The rest of the world has adopted the *International System of Units, SI (Système International)*, a metric system. This system is based on multiples of ten, convenient for folks who still count on their fingers and toes.

Scientists boast that they use uniform, standardized and consistent terminology. That's a suspect claim, however. Consider the use of suffixes such as -scope, -meter, and -graph to describe measuring

The foot-candle.

instruments. "Scope" refers to something one looks through. "Meter" indicates something which has a graduated measuring scale for quantitative measurements. "Graph" designates an instrument which makes a permanent record of measurements. This works fine for the baroscope, barometer, and barograph (an instrument which responds to differences in pressure). It also consistently applies to the spectroscope, spectrometer, and spectrograph, used to analyze the spectrum of wavelengths of light. But the whole system breaks down when applied to the words "telescope," "telegraph," and "telemeter" which have nothing to do with each other. Consider "monocular" and "monograph," or "paragraph" and "parameter." Consistency? Ha!

Creative definitions

Acoustic: The tool used in shooting pool.
Atom: Eve's husband.
Copper nitrate: Overtime pay for policemen.
Erg: An impulse or desire.
Error theory: An excuse for not getting the right answer.
Exit pupil: Student who drops the course.
Faraday: Comes after Thursaday.
Half life: Having to study all of the time.
Helium: What a physician does.
Hydrogen: Put the bottle away.
Ion: What you press pants with.
Joule: Brooklynese for "gem."
Lamina: A female llama.
Liter: A nest of kittens.
Logarithm: A primitive wooden musical instrument.
Lumen: An automobile that gives you trouble.
Momentum: A short moment.
Ohm: Where you live.
Oxide: An ox's outer covering.
Prism: A place where they keep convicts.
Paradox: A partnership of physicians

A logarhythm.

A pair o' docs in a box.

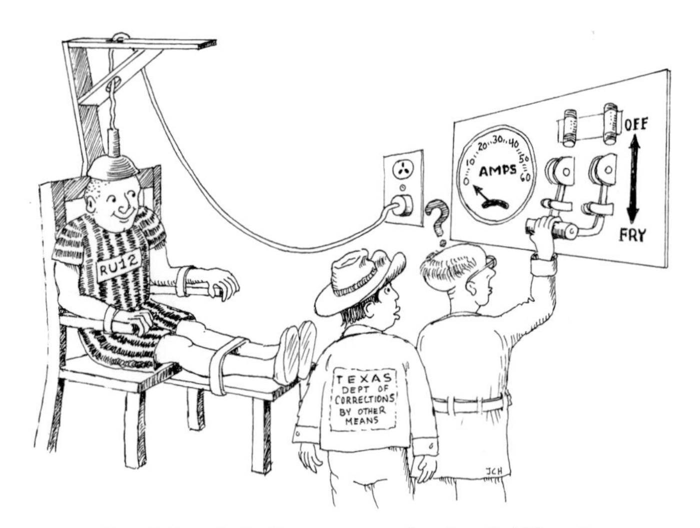

Drat! Somebody forgot to pay the electric bill again.

Seat of emf: An electric chair.
Sin wave: A school of French cinema.
Tension: What the sergeant shouts to his troops.
Virtual work: The useful output of a perpetual motion machine.

One must be careful not to confuse sound-alike technical terms. Determinate-error is one of many kinds of experimental errors one must try to minimize. Determinants are numbers calculated from matrices. Of course, if you do the calculation wrong, it would be a determinant error.

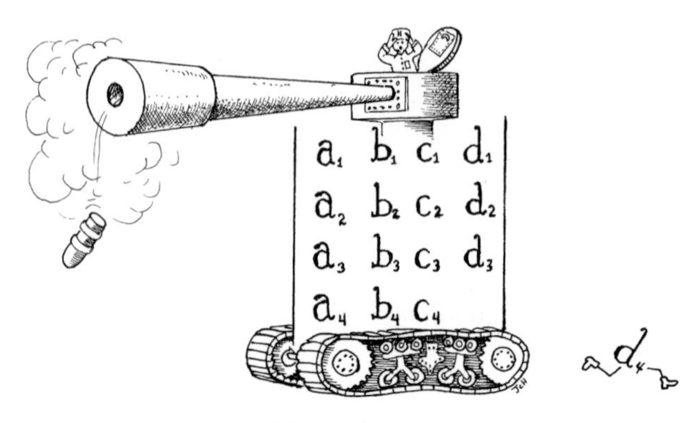

Determinant error.

Metric mayhem

The metric system uses a consistent set of prefixes. Students sometimes have difficulty learning them:

tera = 10^{12}, giga = 10^{9}, mega = 10^{6}, kilo = 10^{3}, deca = 10^{1}
centi = 10^{-2}, milli = 10^{-3}, micro = 10^{-6}, pico = 10^{-12}, atto = 10^{-18}

Perhaps these examples will help:

10^{21} picolos = 1 gigolo
10 millipedes = 1 centipede
10 monologues = 5 dialogues
10^{6} bicycles = 2 megacycles
10^{12} bulls = 1 terabull
10^{1} cards = 1 decacards
10^{-9} goat = 1 nanogoat
1 boo^2 = 1 boo-boo
10^{-12} boos = 1 picoboo
10^{-18} boys = 1 attoboy
10^{12} microphones = 1 megaphone
2 gorics = 1 paragoric
10^{-6} fish = 1 microfiche
10^{12} pins = 1 terapin
10 rations = 1 decoration
$3\frac{1}{3}$ tridents = 1 decadent
2 monograms = 1 diagram
8 nickels = 2 paradigms
2 wharves = 1 paradox

Unitary transforms

Ratio of an igloo's circumference to its diameter: Eskimo pi

2000 pounds of Chinese soup: Won Ton

Time it takes to sail 220 yards at 1 nautical mile per hour: 1 knot-furlong.

365.25 days of drinking low-calorie beer: 1 lite year

16.5 feet measured in the Twilight Zone: 1 Rod Serling

1,000,000 aches or pains: 1 megahurts

Basic unit of laryngitis: 1 hoarsepower

2000 mockingbirds: two kilomockingbirds

1000 grams of wet socks: 1 literhosen

Half the length of a large intestine: 1 semicolon

2.4 statute miles of intravenous surgical tubing at Yale University Hospital: 1 IV League

Now for a pop quiz. What is a demivolt? Answer: Look it up in your Funk and Wagnalls. Yes, it *is* there.

New words and names in science are bad enough, but pronouncing them can be problematical, too. Edmund Halley has a comet named for him. But is the "all" in his surname pronounced as in "ale" or as in "alley"? Diligent historical research suggests that even his family wasn't consistent about this.

A unit of thermal energy is named after James Prescott Joule. Is that pronounced as "jewel" or "jowl"? Here's a poetic answer:

The Joule

You'll be thought cool
If you call it "jewel".
But there'll be a howl
If you call it "jowl".

Planetary embarrassment

Then there's the astronomer's embarrassing dilemma deciding how to pronounce the name of the seventh planet from the sun: Uranus. Either way one might pronounce it sounds scatological, and both pronunciations are sanctioned as "correct" by most dictionaries.

Correct units make a big difference

Students accustomed to inches may have some difficulty with the measurements of a Playboy centerfold expressed in centimeters. The following poem by Harris Stewart illustrates the problem.[1]

---

[1] Reprinted by permission from *The Id of The Squid* by A. E. Benthic, Illus. by J. C. Holden. Compass Publications, 1970. p. 48.

***From A. E. Benthic* The Id of the Squid.**
Compass Publications, 1970, p. 48.

Sentiment and centimeters don't mix

I have some aversion to metric conversion,
Though it's sound from the scientist's view,
But describing a dame will not be the same
When she's 96 - 61 - 92.

Physicist, deep in thought, attempting a proof by induction.

A scientist, making a breakthrough.

A moment of inertia. Perhaps brought on by reading Newton's Principia. From J. C. Holden et al. Now That You Are Here . . . A Guide to the Methow Valley. *The Bookmaker, 1976, p. 40.*

Digital curl

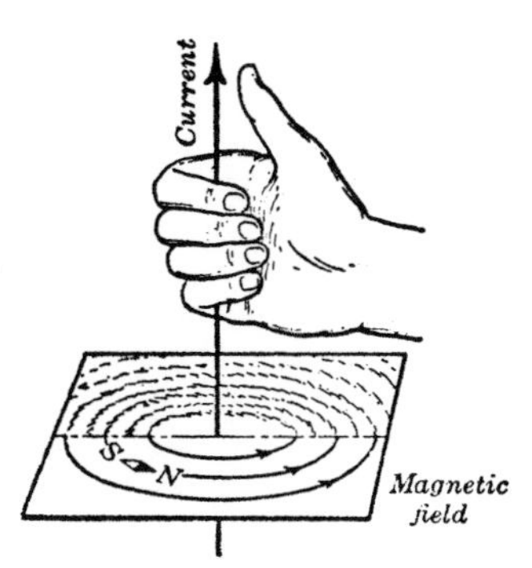

The special relationships describing charges moving in magnetic fields are tricky to remember, so some textbooks suggest various "hand rules" as mnemonic devices. One of the simpler rules states that if you grasp a current-carrying wire with your right hand, the thumb in the direction of the conventional current, then the fingers curl around the wire in the direction of the magnetic field lines produced by that current. That's safe enough if you only imagine it your mind. But don't be tempted to *really* grasp an uninsulated high potential wire in this way if you are grounded.

There's also the "three-finger rule" of induced current. Let the thumb and first two fingers of the right hand be positioned mutually at right angles to each other. Then associate the *F*ield direction with the *F*irst finger, the conventional *C*urrent with the *C*enter finger, and the force (*T*hrust) with the *T*humb. This rule gives the correct direction of the current induced in a wire being moved in a field.

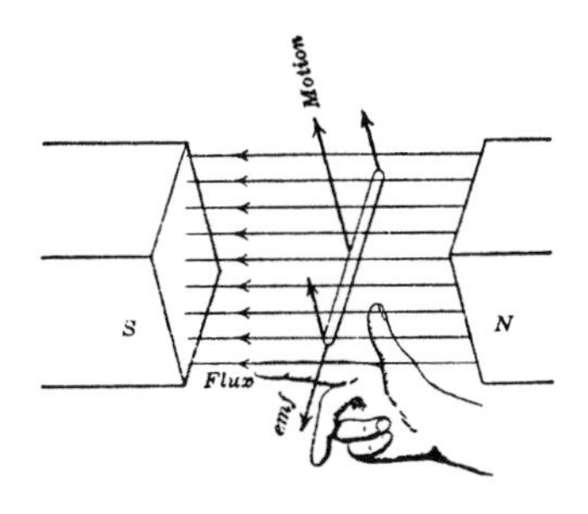

Do not confuse this rule with the motor rule which requires the *left* hand. And if you want to think in terms of *electron* currents (opposite to conventional current) you must switch to the *other* hand in all of these rules. Now if your motor is made of anti-matter... Well, this should be sufficient to make it all perfectly clear.

Students use the hand rules as mental crutches. When they take physics exams you see them engaged in gymnastic hand-waving, trying to position their fingers to the requirements of the problems. Some look as if they were about to sprain a finger.

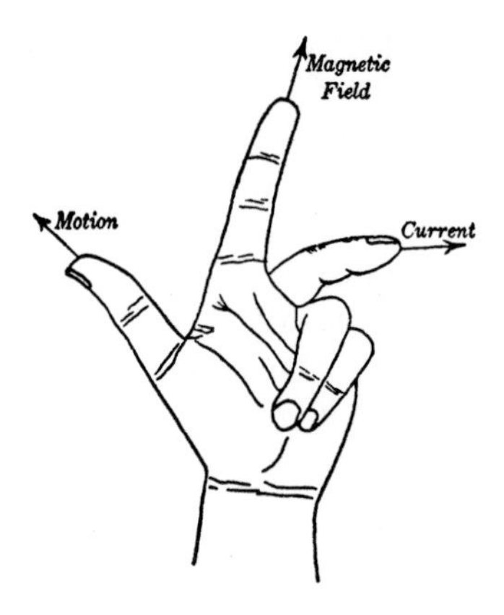

The next picture shows an extreme cases of a malady, *digital curl*, which can result from too much of this sort of exercise. In the interests of classroom safety we warn physics teachers and students that these hand rules can be hazardous.

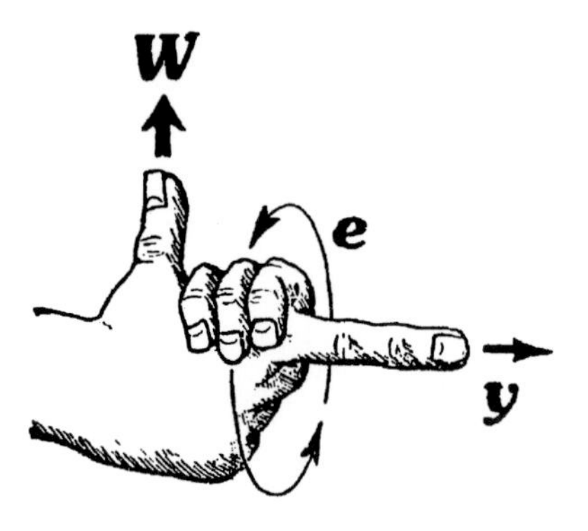

A severe case of digital curl.

These rules can also be hazardous to a student's grade if not used properly. Some students can never remember whether the *I*ndex finger should be associated with current, since *I*ndex begins with *I*, or with *F*ield, since it's the *F*irst finger. Some can't remember whether the *M*iddle finger represents *M*otion, or whether it should be *C*urrent, since it's the *C*enter finger. One girl in a physics class correctly remembered the mnemonic, but always managed to point the middle finger of her right hand forward, the thumb up, and the index finger to the *left*.

There are also unexpected dangers to these rules. Consider the sad case of Philbert Cramm. He was sitting at breakfast in the student cafeteria reviewing his notes for an early morning physics exam. Deep in thought and oblivious to his surroundings he practiced the hand rules. A physical education major sitting across the table took the strange hand motions to be obscene gestures, and punched poor Philbert in the nose.

28
TECHNICAL WRITING

From time to time we see truly incredible examples of the art of technical writing, especially in instruction manuals. The Modesto (CA) [computer] User Group Newsletter is the source of this one:

> The down button, when depressed, will remain down but will not stay up unless the up button has been pressed down. The up button, on the other hand, will not remain down and returns up when pressed down. Should the up button stick down, pressing the down button may possibly cause the up button to release from down and return up. If a situation exists in which the up and down buttons are neither up or down, but the up button is halfway down and the down button is halfway up, then pressing the up button will cause both the up and down buttons to come up.

Most technical publications are not noted for wit and style. It's refreshing when a bit of humor does shows up in one, especially when a publication or corporation is willing to poke a little fun at itself.

The Eastman Kodak company has always done a good job promoting its products through many printed publications. It must be a demanding job to write these; to work up enthusiasm to accurately describe the dry technical details of a product.

A "neutral test card" is a piece of cardboard with a uniform gray surface, manufactured to have a constant, reliable 18% reflectance for light. Professional photographers use it as a standard of comparison, and for taking controlled light meter measurements of studio lighting. Exciting? Hardly! The product is gray and dull. Could it be improved? Glamorized? One of Kodak's creative writers apparently decided to try. The result, datelined April 1, appears in Kodak's publication *Tech Bits*, issue 2, 1984:[1]

> **KODAK NEUTRAL TEST CARD, MARK II**—Research resources and marketing minds have combined to do the impossible—improve that staple photographic apparatus, the Kodak Neutral Test Card, popularly known as the "18 percent gray card." For the brighter world of tomorrow, the Mark II test card now boasts *19 percent reflectance*. With one per package, the card is freshness dated (are you still using last year's card?). In that fashionable neutral gray color and convenient 8 × 10 inch size, the new test card sports many features in common with the current product: child-proof plastic wrap, works with your own meter, solar powered (needs no batteries), same results with either English or metric measurement, use with lenses of all types. For the future, a whole new series of special-purpose test cards is planned. **Kodak Telephoto Neutral Test Card**—same features as the Mark II, but measures 8 × 10 *feet* (not inches). **Kodak Stereo Neutral Test Card**—Same as Mark II, but packaged in squares marked left and right. **Kodak SLR35 Neutral Test Card**—Same as Mark II, but having 2×3 aspect ratio of the popular 35 mm film frame. **Kodak Reverse Neutral Test Card**—Same as Mark II, except that the standard surface is located on the *back* of the card, making it ideal for backlighted scenes. **Kodak Precision Neutral Test Card**—reflectance of 19.000 percent, calibration traceable to the US National Bureau of Standards, neutrality certified by a bureau of the Swiss government, for use only in total darkness (to maintain calibration accuracy), for use with films of highest resolving power, corrected for axial aberrations.

[1] Used with permission.

A remarkable parody of technical writing appeared in the newsletter of the Optical Sciences Center of the University of Arizona. This item sets out to explain, "in simplified terms," the operation of a typical inertial guidance system:

> ... The missile knows where it is at all times. It knows this because it knows where it isn't. By subtracting where it is from where it isn't, or where it isn't from where it is (whichever is greater), it obtains a difference or deviation.
>
> The inertial guidance system uses deviations to generate corrective commands to drive the missile from a position where it is, to a position where it isn't, arriving at the position where it wasn't, but now is. Consequently, the position where it is, is now the position where it wasn't, and it follows that the position where it was is the position where it isn't.
>
> In the event that the position where it is now is not the position where it wasn't, the system has acquired a variation. Variations are the difference between where the missile is and where the missile wasn't and are caused by external factors, the discussion of which is not considered to be within the scope of this report. If variation is considered to be a significant factor, it too may be corrected by the use of a special sub-system. However, the missile must now know where it was also.
>
> The "thought process" of the missile is as follows: because a variation has modified some of the information which the missile has obtained, it is not sure where it is. However, it is sure where it isn't (within reason) and it knows where it was. It now subtracts where it should be from where it wasn't (or vice versa) and by differentiating this from the algebraic difference between where it should be from where it was, it is able to obtain the difference between its deviation and its variation, which is called error...

We'd love to read the rest of the paper from which this was supposedly excerpted. Doesn't this make you feel a lot better about military defense initiatives? Hmm... One wonders how this piece would translate into Russian. The next item may give a clue.

Artificial intelligence

"Artificial intelligence," one of the current buzz words in computer circles, holds forth the promise of really "smart" computers, ones which will know what you mean even when you don't, and perhaps even smart enough to do what you intended instead of what you actually instructed them to do.

We noticed an article "Thinking about Thinking PCs" in *PC World*, Dec, 1986, by Howard Rheingold, which discussed the problems of designing "artificial intelligence" computer programs for language translation.

> The infamous first try at English–Russian translation sent the natural-language computer researchers back to the drawing board. "The spirit is willing but the flesh is weak" was translated into Russian and back into English, yielding "the vodka is agreeable, but the meat has spoiled." Another computer translation transformed the slogan "Coke adds life" into the Chinese sentence "Coke brings your ancestors back from the grave."
>
> How much would a program have to "know" in order to choose one of the two meanings for the sentence: "He saw that gas can explode?"

We have a long way to go. Artificial intelligence (AI) is still very artificial. We have, however, observed more and more examples of AI in our college classes.

What if scientists wrote nursery rhymes

Here are examples of what might result if scientists re-wrote children's nursery rhymes. See how many you recognize. Answers are below.

1. A research team proceeded toward the apex of a natural geologic protuberance, the purpose of their expedition being the procurement of a sample of fluid hydride of oxygen in a large vessel, the exact size of which was unspecified. One member of the team precipitantly descended, sustaining severe fractural damage to the upper cranial portion of his anatomical structure.

Subsequently, the second member of the team performed a self-rotational translation oriented in the direction taken by the first member.

2. Complications arose during an investigation of dietary influence: one researcher was unable to assimilate adipose tissue and another was unable to consume tissue consisting chiefly of muscle fiber. By reciprocal arrangement between the two researchers, total consumption of the viands under consideration was achieved, thus leaving the original container of the viands devoid of contents.

3. A young male human was situated near the intersection of two supporting structural elements at right angles to each other: said subject was involved in ingesting a saccharine composition prepared in conjunction with the ritual observance of an annual fixed-day religious festival. Insertion into the saccharine composition of the opposable digit of his forelimb was followed by removal of a drupe of genus prune. Subsequently the subject made a declarative statement regarding the high quality of his character as a young male human.

4. A triumvirate of murine rodents totally devoid of ophthalmic acuity were observed in a state of rapid locomotion in pursuit of an agriculturalist's marital adjunct. Said adjunct then performed triple caudectomy utilizing an acutely honed bladed instrument generally used for the subdivision of edible tissue.

5. A female of the species *Homo sapiens* was the possessor of a small immature ruminant of the genus *Ovis*, the outermost covering of which reflected all wavelengths of visible light with a luminosity equal to that mass of naturally occurring microscopically crystalline water. Regardless of the translational pathway chosen by the *Homo sapiens*, the probability was 1 that the aforementioned ruminent would select the same pathway.

6. A human female, extremely captious and given to opposed behavior, was questioned as to the dynamic state of her cultivated tract of land used for production of various types of flora. The tract components were enumerated as argentous tone-producing agents, a rare species of oceanic growth and pulchritudinous young females situated in a linear orientation.

Answers

1. Jack and Jill went up the hill
   To fetch a pail of water.
   Jack fell down and broke his crown,
   And Jill came tumbling after.

2. Jack Sprat could eat no fat.
   His wife could eat no lean.
   And so between the two of them,
   They ate the platter clean.

3. Little Jack Horner
   Sat in the corner
   Eating his Christmas pie.
   He stuck in his thumb
   And pulled out a plum
   And said "What a good boy am I!"

4. Three blind mice, three blind mice
   See how they run, see how they run.
   They all ran after the farmer's wife
   Who cut off their tails with a carving knife.
   Did you ever see such a sight in your life
   As three blind mice?

5. Mary had a Little Lamb
   Whose fleece was white as snow.
   And everywhere that Mary went,
   The lamb was sure to go.

6. Mary, Mary, quite contrary,
   How does your garden grow?
   With silver bells, and cockle shells
   And pretty maidens, all in a row.

What if Dr. Seuss wrote technical manuals?

If a packet hits a pocket on a socket on a port,
And the bus is interrupted as a very last resort,
And the address of the memory makes your floppy disk abort,
Then the socket packet pocket has an error to report!

If your cursor finds a menu item followed by a dash,
And the double-clicking icons put your window in the trash,
And your data is corrupted 'cause the index doesn't hash,
Then your situation's hopeless, and your system's gonna crash!

If the label on your cable on the gable at your house,
Says the network is connected to the button on your mouse,
But your packets want to tunnel to another protocol,
That's repeatedly rejected by the printer down the hall.

And your screen is all distorted by the side effects of gauss,
So your icons in the window are as wavy as a souse,
Then you may as well reboot and go out with a bang,
'Cause as sure as I'm a poet, the sucker's gonna hang!

When the copy of your floppy's getting sloppy on the disk,
And the microcode instructions cause unnecessary RISC,
Then you have to flash your memory and you'll want to RAM
    your ROM,
Quickly turn off your computer and be sure to tell your mom!

29

ANON
THE MYTH BEHIND THE LEGEND

by DONALD E. SIMANEK[1]

Few literary puzzles have inspired such universal apathy as the question: "Who was Anon?" Books of quotations are cluttered with sayings attributed to Anon, and these scraps of truth and wisdom have earned Anon universal recognition and immortality. Innumerable biographies have been written about lesser authors, even authors so obscure that their works are seldom read. But Anon himself, though widely read and widely quoted, has been accorded only widespread indifference by the literary community.

Anon. Artist's conception based on various unreliable sources.

Even the most astute literary scholar would be perplexed if asked to identify the central themes of Anon's work. If a historian

[1] From *The Vector*, **7**, 2 (May, 1983) p. 18–20.

were asked how Anon's work was influenced by the culture and events of his times, he would be at a total loss for a sensible answer.

So complete has been the scholarly neglect of Anon that his name has become a synonym for "unknown." In spite of this, his works have stood the test of time, and he continues to be one of the most often quoted authors. (Ibid may be more frequently cited, but his works were derivative.)

What little we know of Anon's life is of doubtful validity. We have no authentic picture of him, nor any firsthand description of him by anyone who would admit having known him. Not one scrap of original manuscript in his own hand has survived the ravages of time. Scholars have given up hope of ever discovering an autobiography of Anon in some dusty library.

Yet, from the available dearth of evidence, we can piece together a sketch (albeit apocryphal) of this prolific genius. We know that Anon's wisdom appeared very early in history. When references to him are traced backward in time, in the general direction towards the emergence of civilization, they lead us to a blank wall. This suggests that Anon must be placed in historical times so ancient as to predate the emergence of intelligent thought. He was certainly ahead of his time, which may be the reason why none of his contemporaries knew of him.

If that argument seems insufficient, consider this independent and equally convincing evidence which leads to the same conclusion. Anon's work was considered immortal in all historical ages, and it is generally quite difficult for an author to achieve immortality in his own time.

Perhaps Anon inspired an ancient "school" of thinkers who later traveled far and wide disseminating his ideas. This may be true. Nobody knows. But then, he would, since Nobody knew Anon personally. Indeed, Nobody knew a lot of things which baffled everyone else. But the hypothesis that Nobody was a pupil of Anon is dubious, if true.

The historical problem is compounded by the timeless quality of Anon's work. His wisdom seems too old-fashioned for modern times, yet too advanced for ancient times. Either Anon was in the habit of living in the past, or anticipating the future.[2] If so, it follows

[2] This may be the origin of our word "anon" meaning "soon".

that he was probably neglected and unappreciated in his own age, and that could explain a lot.

Leaving these irrelevant questions aside, let us look at Anon's career. It can be divided into three distinct phases: the first, the second, and the third. That leaves only the problem of deciding into which phase to place each of Anon's works. This is especially troublesome for his posthumous works. Since we have no idea when Anon died, it's even a bit difficult to determine *which* of his works *were* posthumous.[3]

We might at least hope to extract Anon's philosophy from those fragments of his genius which have trickled down to us through the sieve of history. It is a vain hope. While Anon wrote (or perhaps spoke) on many subjects, he had the infuriating habit of speaking on every side of every question. No consistent pattern emerges, but this is itself consistent with Anon's own observation that "Consistency is the curse of small minds." On yet another occasion he said, "Sticking consistently to any one position sooner or later leads to logical difficulties." Perhaps Anon merely wanted to ensure that all sides of every question be heard. Yet he expressed reservations about this approach, saying, "One who can see both sides of a question doesn't understand the question." Such remarks strongly suggest that Anon may be the true father of the disciplines of logic and philosophy.

To achieve a true appreciation of Anon's work we must first recognize that the inconsistencies and contradictions inherent or implied in his work do, in fact, represent the central, unifying theme of his philosophy.

Anon's fragmentary output has become so diffused throughout many cultures that it is nearly impossible to specify his country of origin. Some have suggested that Anon was German, his full name being Till Anon—a ridiculous notion at best. Another improbable theory has it that Anon was Spanish with a German surname: Anon y' Maus. Or could this be a nickname describing Anon's timidity: "Anon, the mouse"?

One historian even goes so far as to suggest that all of Anon's works are forgeries of recent (19th century) origin, perpetrated by

[3] Since we don't know when Anon was born, some of Anon's works might have to be classified as pre-natal.

author Lewis Carroll (Charles Dodgson) writing under the pseudonym: E. M. Anon. When this name is read backwards it is seen to be an anagram of the kind Carroll loved to devise. This outrageous theory deserves to be rejected on its merits.

Anon, from an old print of questionable authenticity.

Lest we be overawed by Anon's versatility, we should look at what he *didn't* do, for that demonstrates his discrimination and good taste. He never wrote an epic poem, a play, or an opera. He never wrote a bestselling work of fiction, never wrote a textbook, and never edited an anthology. He left such enterprises to hacks and lesser intellects.

No painting or drawing bears the signature "Anon." No sculpture has "Anon" chiseled on its base. If Anon ever tried his hand at art, he apparently never signed his works.

For all of his output of serious sagacity, homely homilies, and profound pronouncements, Anon had a lighter side. In fact his output of jokes far exceeded the rest of his literary work. It is true that many of these jokes are off-color, but that has only enhanced their popularity. They are remembered and quoted verbatim by people who couldn't recite one line of "The Ancient Mariner." Anon knew that art is of no value without an audience, or as he put it so well, " 'Tis better to be obscene than unheard."

So, a picture of Anon emerges: a witty, slightly cynical, philosopher of the people. He could sum up the essence of an idea in one pithy sentence. Though many others plagiarized his works, he never complained. He must have cared little for money, for there is no record that he was ever paid for any of his work.

Anon demonstrated that the best way to achieve recognition is by not seeking it. He was unconcerned about the judgment of posterity, for he said, "Be not obligated to posterity. What has posterity ever done for you? In any case, the critical judgments of posterity come too late to be useful."

Of course any conclusions about Anon, the man, might have to be modified if it were shown that Anon was a woman. The true sex of Anon may be a matter of dispute among scholars, yet we have no reason to believe that Anon ever had the slightest concern about this question.

As usual, Anon had the first word on such speculations when he (or she) said, "Nothing stimulates outrageous theories so effectively as an absence of evidence."

Appendix: scholarly notes

1. The name "Anon" is virtually unknown in any language, which suggests that Anon had no descendants. Perhaps Anon's family suffered from hereditary infertility. It's a well known biological fact that if your parents had no children, it's very likely that you won't either.

2. Recently Anon's works have been subjected to stylistic analysis with the aid of a computer. The tentative conclusion is that Anon plagiarized all of his works from others.

3. Those who fault Anon's style should remember that his sayings would probably sound better in Anon's native tongue, if we only knew what language that was.

4. We may safely assume that Anon never had the advantage of higher education, for no Ph.D. thesis bears his name.

5. Though Anon's life is shrouded in obscurity, his works have far greater merit than those of authors whose *meaning* is shrouded in obscurity.

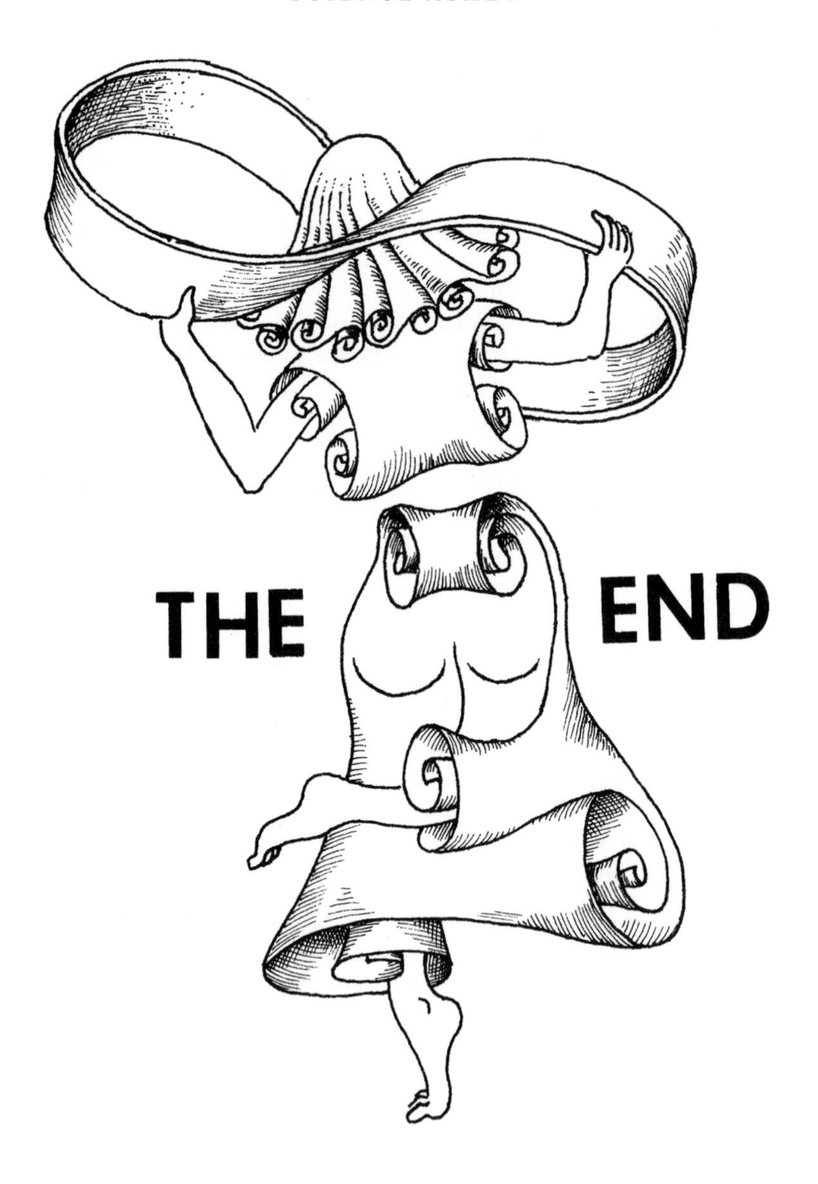
THE
END